BAYOU-DIVERSITY

BAYOU-DIVERSITY

Nature and People in
the Louisiana Bayou Country

Kelby Ouchley

Louisiana State University Press
Baton Rouge

QH
105
.L8
O93
2011

Published by Louisiana State University Press
Copyright © 2011 by Louisiana State University Press
All rights reserved
Manufactured in the United States of America
First printing

Designer: Laura Roubique Gleason
Typeface: Minion Pro
Printer: McNaughton & Gunn
Binder: Acme Bookbinding

LIBRARY OF CONGRESS CATALOGING-IN-PUBLICATION DATA

Ouchley, Kelby, 1951–
    Bayou-diversity : nature and people in the Louisiana bayou country / Kelby
Ouchley.
        p. cm.
    ISBN 978-0-8071-3859-5 (cloth : alk. paper) — ISBN 978-0-8071-3860-1 (pdf)
— ISBN 978-0-8071-3861-8 (epub) — ISBN 978-0-8071-3862-5 (mobi)
    1. Biodiversity—Louisiana. 2. Bayous—Louisiana. 3. Species diversity—Louisi-
ana. I. Title. II. Title: Bayou diversity. III. Title: Nature and people in the Louisiana bayou country.
    QH105.L8O93 2012
    578.76809763—dc22
                                    2011008225

The following essays first appeared in print in *Big Muddy: A Journal of the Mississippi River Valley:* "Coming Back," "Swamp Snow," and "Tributary Affair."

To my sons, Jeremy and Zachary,
my connections to the natural world of the future

CONTENTS

2. Bayou-Fauna: Animals in a Place of Bayous   40

3. Modi Operandi: Methods of Functioning or Operating 118

CONFLUENCE: A Flowing Together of Two or More Streams of Life in a Physical Place

4. Encounter: A Meeting, Especially One That Is Unplanned, Unexpected, or of Note 151

ACKNOWLEDGMENTS

Since 1995 the Monroe, Louisiana, public radio station, KEDM 90.3 FM, has aired my short biweekly radio program entitled *Bayou-Diversity*. Many of the hundreds of narrated essays written for that program appear in this book. I am happily indebted to past and present employees of this great station who have enabled my work over the years, especially Ray Davidson.

My publishing adventure with LSU Press continues to be a journey of positive experiences. Accordingly, I thank MaryKatherine Callaway, Rand Dotson, Lee Sioles, Erin Rolfs, Michelle Neustrom, and Bobby Keane for guiding me forward. I also appreciate freelance copyeditor Maria denBoer for her cheerful and precise finetuning of the manuscript.

Emily Caldwell, the illustrator for this book, doesn't realize the vast extent of her potential as an artist. I, among many others, do.

Imagine the dinner table talk of two naturalists who have been married almost forty years. The wellspring of many of these stories and all of the joy in my life flow from Amy.

BAYOU-DIVERSITY

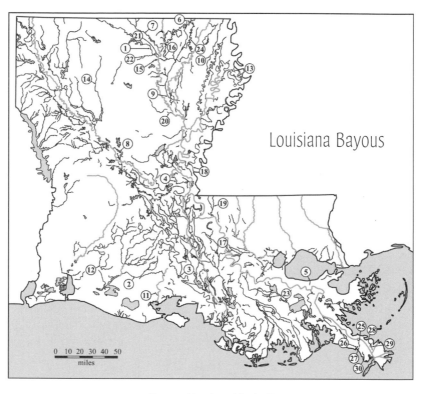

Louisiana Bayous

Bayous Mentioned in the Text

1. Bayou D'Arbonne
2. Bayou Queue de Tortue
3. Bayou Teche
4. Bayou Jeansonne
5. Bayou St. John
6. Bayou Bartholomew
7. Bayou de L'Outre
8. Bayou Rigolette
9. Bayou Lafourche
10. Bayou Macon

11. Little Bayou
12. Lacassine Bayou
13. Walnut Bayou
14. Saline Bayou
15. Tupawek Bayou
16. Bayou de Siard
17. Elbow Bayou
18. Bayou Cocodrie
19. Bayou Sara
20. Bayou Dan

21. Cross Bayou
22. Bayou Choudrant
23. Bayou Boeuf
24. Bayou Bonne Idee
25. Bayou Tambour
26. Bayou Little Channel
27. Bayou Tony
28. Buras Bayou
29. Long Island Bayou
30. Felice Bayou

INTRODUCTION

I have been blessed by spending most of my life near bayous. At various times I have lived within a few miles of Bayou D'Arbonne, Bayou Queue de Tortue, Bayou Teche, Bayou Jeansonne, Bayou St. John, Bayou Bartholomew, Bayou de L'Outre, Bayou Rigolette, Bayou Lafourche, Bayou Macon, Little Bayou, Lacassine Bayou, Walnut Bayou, Saline Bayou, and Tupawek Bayou. I obtained a college degree in wildlife biology from a university on the bank of Bayou de Siard and pined constantly for bayous while taking a second degree in a land where mesquite replaced cypresses. As a biologist and manager of National Wildlife Refuges for thirty years, I received a salary for work on bayous and their watersheds. Whether pursuing poachers down a bayou at midnight or writing land protection plans for their swamps, it was a privilege that never ceases to amaze me. I chose as my wife a woman who was raised in a home 100 feet from a bayou. The two sons born to us grew to manhood near bayous. For the last twenty years I have lived on the edge of the D'Arbonne Swamp, bayou backwaters of spring floods only a rock's throw from my cypress house. A nearby low ridge adjacent a cypress brake was once an encampment for prehistoric people of the Coles Creek culture. At the time a bayou with a name that vanished with their language flowed hard against it. Seven hundred years later, more or less, my great-grandfather plowed cotton behind a petulant mule in the rich midden soil. Someday the ridge will receive my ashes, eventually recycling a few on down into Bayou D'Arbonne.

When the time seemed right to compile these short essays, bayous in a broad sense presented a unifying theme. The collection began in 1995 as narrated vignettes for a weekly public radio program called *Bayou-Diversity*. It is important to know that each story in this book was selected because it has a connection to bayous, although for some the nexus may be floating just beneath the surface. A primary maxim of modern environmental education espouses the connectivity of all living things and the dangers of assuming that humans can be separated from the natural world without consequences, even in thought. For me the book will only be a success if it stimulates readers to consider how a white oak acorn on the highest hill in Louisiana, a cell phone tower, a disturbed man's ranting, and a natural gas well are connected to bayous.

Visitors to Louisiana are often perplexed by the peculiar terms we use to describe our natural features, such as sloughs, swamps, and brakes. One of the most common questions asked by naïve outsiders is, "What is a bayou?" The word is ubiquitous here; after all, this is the Bayou State. The confusion is understandable, as a broad range of businesses in the state advertise the word "bayou" in their names. There is a Bayou Bowling Alley, Bayou Builders, Bayou Forklifts, Bayou Gymnastics, Bayou Internet, Bayou Plumbing, and so forth on down through the alphabet. We have "bayou" churches, schools, and of course the fighting Bayou Bengal football team.

Early Choctaw Indians would be mystified at all this hoopla. They were responsible for the etymology of the word. French settlers took the Native Americans' perfectly good word "bayuk" and contorted it into the Franco-label "bayou." The definition, however, remains the same. It is a natural, relatively small waterway that flows through swamps and other lowlands in most cases. Except during flood events, currents in bayous are usually sluggish or absent. Crystalline waters are not a characteristic of bayous, as they meander through heavy clay soils and capture the washed-in sediments of subtropical rains. Their shores are often dressed in live oaks and big-butted cypress trees laden with Spanish moss and parula warblers. Other riparian areas comprise agricultural fields of cotton, soybeans, rice, and sugar cane, or infrequently pines and upland hardwoods. Inhabitants of bayous have notable reputations and include crawfish, cottonmouths, mosquitoes, alligator gars, alligator snapping turtles, and alligators. Plenty of other species, more innocuous in their manners, are dependent on bayous and their watersheds.

Even many Louisiana citizens are not aware of the extent that bayous braid the state from north to south. More than four hundred named bayous in sixty-two of sixty-four parishes seek the shortest route to the Gulf of Mexico across a landscape void of significant relief. A few creep into the flat fringes of other Gulf Coast states and Arkansas, where Bayou Bartholomew, the longest bayou in the country (about 375 miles), heads up. Bayous exist in the state's largest city of New Orleans (Bayou St. John), in the smallest village of Mound (Walnut Bayou), and within the corporate limits of the capital of Baton Rouge (Elbow Bayou). Glorious places if undammed, unditched, and unpolluted, they are worthy of conservation if for no other reason than Louisiana without the living gumbo of bayous would not be Louisiana, the Bayou State.

Biodiversity can be defined as all the varieties of life forms in a certain area. The area can be as large as planet Earth, where an estimated 10 million species of plants, animals, and microbes live (95 percent of which are made up of arthropods and microbes) or as small as a single drop of bayou water. Although the diversity of life is sometimes viewed at levels as minute as individual genes, it is more commonly considered at the scale of ecosystems. Within Louisiana several broad ecosystems are delineated by dominant vegetation types and include coastal marshes, bottomland hardwoods, prairies, pine forests, and mixed pine/upland hardwood forests. Each of these categories can be divided into more refined classifications (e.g., saline, brackish, and freshwater marshes).

Species richness denotes the number of different species of

plants, animals, and microbes in a given area. Different ecosystems vary in their natural capacity to support different types of life. In general, biodiversity decreases as one moves farther from the equator or higher in elevation. An important indicator of an ecosystem's health is derived by comparing current species richness against what might be expected in the area if unaltered by human disturbance. As an example, a dredged bayou draining a polluted swamp would likely have low species richness and thus poorer biological health compared to a free-flowing bayou with a pristine watershed.

The health of human societies depends on ecosystems that are species rich by supporting processes that provide benefits to everyone. Air and water are filtered. Climate changes are moderated when forests sequester carbon dioxide. Wetlands mitigate the impacts of hurricanes and store floodwaters that could otherwise be devastating. Critical agricultural benefits are amassed in the genetic traits found in wild varieties of domestic crops. Likewise, human health benefits accrue with biodiversity, as many drugs are derived, directly or indirectly, from biological origins. At least 50 percent of pharmaceuticals in the U.S. market stem from natural compounds found in plants, animals, and microorganisms. Fish, seafood, and some species of native plants and wildlife are biodiversity components that provide food and recreational opportunities. For many people the aesthetic and spiritual values of intact biodiversity are vital and immeasurable.

Bayous + Biodiversity = Bayou-Diversity. It is the variety of all living things in a place of bayous with their integral watersheds that encompass the whole of Louisiana. To a greater extent, it is the stories of every affair of natural history in the state—the stories of flora, those that produce seeds, spores, and rashes; the stories of fauna, feathered, scaled, furred, and finned. It is stories of their environment. *Bayou-Diversity* is also stories of human encounters with these wild things and wild places, enriching and surprising, troubling and promising.

BIOTA

The Plant and Animal Life of a Particular Region

1
BAYOU-FLORA
Plants in a Place of Bayous

Spent day w/Dr. Thomas on refuge [near Judd Bayou]; found rare shell-
bark hickory grafted on pecan stump, 20" dbh sassafras, & black gum.
    —KO Field Diary, 27 September 1983—Tensas River NWR

To Riverwood "protected" tract in Caldwell Parish near Cypress Creek
to check old known site of yellow ladies slipper; has not been seen here in
several years since adjacent clearcut—Found 3 plants, not in flower yet,
leaves about 5" long; I've never seen this plant in La. before.
    —KO Field Diary, 2 April 1996

No less than 3,249 species and subspecies of vascular plants have
been found growing in Louisiana. Of this number, 826 (25 percent)
are considered non-native in the state and were introduced intentionally or accidentally. Many in this category are troublesome invasive species, which displace native plants, reduce species richness,
and disrupt ecosystem processes. Chinese tallow and water hyacinth are examples. The native plants include generalists, like deciduous holly, which grow in a wide variety of habitats and are found
in every parish, and specialists with exacting requirements, such
as leatherwood, which grow only in restricted niches. Their stories
abound.

Baldcypress I

Nothing characterizes a southern swamp more than a giant, moss-draped cypress tree standing knee-deep in a backwater slough. Technically known as baldcypress, these survivors of ancient life forms once found across North America and Europe are now greatly restricted in range. In the United States they are native to river bottoms and swamps in the Deep South and along the eastern seaboard north to Delaware. In Louisiana, although the last large virgin stands are gone, cypresses can still be found in every parish.

Cypress trees once grew to 17 feet in diameter and 140 feet in

Baldcypress

height. They were the largest trees in the South and lived to be 400 to 600 years old. A few were estimated to be more than 1,000 years old. Even though cypresses are at home in wetlands, their seeds cannot germinate underwater and young seedlings soon die if they are overtopped by floodwaters during the growing season. For that reason, the trees growing in Monroe's Bayou DeSiard, Old River in Natchitoches, and Lake Pontchartrain near New Orleans began life on dry or muddy ground that was not flooded during the growing season for at least a couple of years. Older trees can adapt to intermittent flooding regimes and usually develop fluted trunks, but permanent, deep flooding will eventually kill most mature trees. Throughout the state, including the areas mentioned above, a steady decline of cypresses resulting from human-altered flooding regimes is ongoing.

Historically, cypress has always been important to humans in Louisiana. The reddish heartwood of old trees is durable and resistant to decay in a climate that fosters the rapid decomposition of most woods. For thousands of years Native Americans in bayou country used cypress for dugout canoes. Early colonists were quick to discover its value as a building material. In 1797, Don Juan Filhiol described Fort Miro, the first sizable colonial structure in the Ouachita Valley, as "an enclosure in post of tipped cypress . . . in an area [in]which is found the principal house . . . covered in cypress shingles." In the late 1800s, the demand for cypress lumber for boats, furniture, pilings, trim, shingles, siding, and coffins was great. It was during this period that the vast virgin stands were logged over. By 1925, the once-thriving cypress industry was in a spiraling decline as the last of the raw product was exhausted.

Most cypress stands today are second growth, but there still remain a few giants among us, towering 100 feet above the earth. They exist because they are hollow and thus not merchantable or because they grow in an area so remote as to make harvest unfeasible. They laid down their first annular rings during the classical period of the Mayan culture. They germinated and grew into seedlings as Charlemagne was crowned Holy Roman Emperor. They were sound and mature when the sun gleamed from the swords of Hernando de Soto's men as they marched across northeastern Louisiana in a fruitless search for gold. It is possible that their limbs were once laden with the weight of a thousand passenger pigeons and that their bark was probed by ivory-billed woodpeckers. Cougars and bears may have sought refuge in their hollows. It is likely, too, that a few of these will still be greeting each spring with a fresh feathering of needle-like leaves in centuries to come.

Baldcypress II

In Louisiana cypress trees have their own unique and proper place in the world. For those of us who live among the swamps, that place is along the bayous, lakes, and brakes of our wetlands. Nothing could be more enigmatic than to find this tree growing at the base of steep hills and limestone cliffs adjacent to streams that are crystal clear and actually flowing. Such is the case, however, in the central Texas hill country. It is the last place I would expect to find a baldcypress tree.

For whatever reason they thrive in this environment of drought and sudden flashfloods, leaning downstream like contorted gnomes, often with flood debris 30 feet up in their limbs. Cypress knees are uncommon in this habitat, but their roots form serpentine lattices, some like a ball of giant cottonmouths wrapped around small boulders. Like inverted Medusas they cling to the earth and brace for Mother Nature's tempests. Their cohorts are sycamores and occasional wild sweet pecans.

Up the narrow, *V*-shaped side canyons grow the perfect cypress trees—the tall, straight, tight-barked individuals that escaped the cross-cut saw only because of their inaccessibility. We knew the last of those trees in Louisiana soon after 1900. In Texas, the surviving Amazonians guard transparent dripping pools smothered in maidenhair fern and tower over the nearby live oaks and juniper. Their attendant epiphytes are ball moss instead of the familiar Spanish moss. In their own way they are just as much at home as those along Bayou Teche or in the Atchafalaya Swamp. It is only our perception that sees them as out of place.

Baldcypress III

The legacy of Louisiana's official state tree should not end in your flower bed. Mulching to control weeds and conserve water is a great idea; using mulch made from cypress trees is not. Most mature cypress forests in this state were cut for lumber decades ago. Almost all of the remaining stands are relatively young—too young to be used for timber and too young to reproduce naturally. When clearcutting of these young stands occurs to produce cypress mulch, any chance of future reproduction is eliminated, and, unlike pines, cypress is rarely replanted behind a harvesting operation. The bottom line is that the current rate of baldcypress harvest to fuel the demand for mulch is not sustainable.

Florida was first to recognize this issue, and many counties there restrict the use of cypress mulch. In Louisiana, Governor Kathleen Blanco commissioned a special study entitled *Report to the Governor from the Science Working Group on Coastal Wetland Forest Conservation and Use.* Baldcypress and water tupelo were determined to be the primary species in coastal swamp forests, a critical component of Louisiana's imperiled wetlands. Significantly, the study found a current lack of regeneration in our remaining cypress forests.

Recent research has shown that the popular demand for cypress mulch is at least partly driven by myths. Consumers often buy cypress mulch under the assumption that it is more durable and long lasting. This is not the case since today's mulch is made from young trees yet to develop rot-resistant heartwood. Work by the Florida Co-op Extension Service found other problems: "When dry, cypress mulch repels water, making it difficult to wet, particularly if it is on a mound or slope." Moreover, once it is wet "cypress mulch appears to have a high water-holding capacity that may reduce the amount of water reaching the plant root zone." Even its attractive color soon fades away.

Cheaper and effective alternative mulches are available. A University of Florida study found that wood chips, pine bark, and pine straw rated just as high as cypress. Efforts to stem the loss of Louisiana's coastal wetlands are hollow without a strategy to keep our state tree in the swamps instead of in flower beds.

Baldcypress resides with cohorts that also evoke images of bayou country.

Spanish Moss

Spanish moss is not. What I mean is that Spanish moss is not Spanish and is not a moss. It does not grow in Spain but rather in the southeastern United States down into South America. It is not a true moss like sphagnum but rather a flowering plant in the bromeliad family very closely kin to pineapples. Often associated with our images of southern swamps, Spanish moss grows in long, draping, thread-like, gray veils on trees, where it absorbs moisture and nutrients from the air. The plants are not parasitic and don't harm their host trees.

Many types of wildlife use Spanish moss in their lifecycles. Squirrels and birds use it for nest materials. Along bayous, parula warblers build their nests almost exclusively in draping clumps of the plant. Some species of bats roost in Spanish moss, and it is the sole habitat for one kind of jumping spider.

Humans have used Spanish moss for centuries. Early European colonists recorded Native Americans wearing clothing made from the plant. Louisiana Cajuns made a concoction of mud and Spanish moss known as bousillage for mortar and house insulation. Later, an entire commercial industry developed around the harvest and processing of the plant into manufactured products. It was used for packing materials, as mulch, and in saddle blankets. Thousands of tons were ginned for mattress stuffing until as late as 1975, when synthetic fibers replaced the natural filaments. In recent years, researchers have studied components of Spanish moss as a possible drug to control blood pressure.

Because Spanish moss receives all of its nutrients from the air, it is very sensitive to wind-borne pollutants, such as heavy metals from exhaust fumes and pesticides. Early explorers in Louisiana often remarked about the dismal, dreary atmosphere associated with moss-laden swamps. We now know that the presence of healthy Spanish moss is an indicator of good air quality, and is thus a welcome part of our bayou scenery.

Cane

Canebrakes were once one of the most unique features of the Louisiana landscape. President Teddy Roosevelt described them well in a turn-of-the-twentieth-century visit to East Carroll and Madison parishes. He wrote, "The canebrakes stretch along the slight rises of ground, often extending for miles, forming one of the most striking and interesting features of the country. They choke out other growths, the feathery, graceful canes standing in ranks, tall, slender, serried, each but a few inches from his brother, and springing to a height of fifteen or twenty feet. They look like bamboos; they are well nigh impenetrable to a man on horseback; even on foot they make difficult walking unless free use is made of the heavy bush-knife. It is impossible to see through them for more than fifteen or twenty paces, and often for not half that distance. Bears make their lairs in them, and they are the refuge for hunted things."

In describing the Morehouse Parish haunts of legendary woodsman Ben Lilly, J. Frank Dobie wrote, "Canebrakes stretched for miles and miles, the hollow stalks that waved their green blades fifteen or twenty feet up in the air rooted so densely that only bears, razorback hogs and a man with a knife could penetrate them." He also stated that they were the last refuges for hunted things, including some men.

Commonly known as switch cane, this perennial, woody member of the grass family is found in every parish of the state except those bordering the coast. More than a thousand related species grow mainly in the tropical and subtropical areas of the world and are usually referred to as bamboo. Several of the exotic species have been introduced into Louisiana. One striking characteristic of most species is that they produce seeds only once in their lifetime, which may exceed 100 years, and then die.

For local Native Americans canebrakes furnished abundant raw materials for a plethora of products. It was used in the manufacture of blowguns, darts, arrow shafts, shields, knives, spears, duck calls, whistles, and flutes. Cane was weaved into rafts, baskets, bedding, roofing, and floor and wall coverings. Cane containers held everything from seeds for the winter larder to bones of the dead. Later, European settlers valued cane for cattle forage.

Today only scattered small remnants of the once vast canebrakes can be found. Several factors seemed to have been involved in their decline. Overgrazing by domestic cattle was likely an early influence. Occasional fires and floods are thought to be important requirements in the natural lifecycle of cane. Fire protection, levees, and drainage have decreased the frequency of these events. Without a doubt, though, most canebrakes were converted to agricultural fields. In a voyage to Louisiana in 1803, C. C. Robin wrote, "This reed only grows on land that is never (or almost never) flooded . . . These cane brakes, on account of the large amount of humus that they deposit, make the soil very fertile, and the farmers regard their cane brakes as the best possible land; in fact, they judge the quality of the soil by the thickness of the cane."

Animals around the world are dependent on bamboo for their existence. Endangered pandas feed almost exclusively on bamboo in China, and some birds in Central America are found only in bamboo thickets. Such a relationship may have existed in Louisiana with the Bachman's warbler, a rare or possibly extinct songbird. No documented sightings of this bird have occurred in several years, and most observations in the past have been in association with canebrakes. This example reinforces the connectivity of all living things within our ecosystems. Rarely do we modify even one component of our surroundings without impacting others.

Mayhaw

For those of us who might be considered unrefined epicureans, May is the month of ritual pleasures involving a wild gourmet treat. It is the season to gather mayhaw fruits and make one of the finest jellies to grace a buttermilk biscuit.

Born of southern swamps, mayhaws are small trees technically considered hawthorns in the rose family. They grow in wetlands across the Southeast and are usually found only in soils that have a sandy component. Accordingly they are rare in the heavy clay soils near the Mississippi River and common along the Ouachita/Black River system and its tributaries.

The white mayhaw flowers occur in February and March and often present the first splash of spring color to local woodlands. Flowers usually occur before and during the emergence of leaves. Marble-size reddish fruits resemble small apples and ripen in May and June. An old axiom claims, "If mayhaws flower in the water, they will fall in the water." This refers to the backwater flooding common to most mayhaw habitat. Studies have shown that trees standing in water have a delayed bloom period.

Mayhaws are an important food source for many kinds of wildlife. Deer, raccoons, squirrels, opossums, and several species of birds relish the fruits. Native Americans undoubtedly consumed them for thousands of years, and the first Europeans quickly learned of their value. One pioneer Louisiana diary account reveals that mayhaw-gathering could be quite an adventure. Miss Caroline Poole, a schoolteacher in the frontier village of Monroe, writes in her entry of May 7, 1836, "Hunt for May-haws. Rode sixteen miles on horseback. Saw rattlesnake. Crossed bayous where the water was above the saddle skirts, thirty yards wide. Saw black snakes in abundance. Camped in the woods. Coffee. Bacon cooked on a stick. Enjoyed the day but very much fatigued." A note in *The Gazette* of Farmerville on May 2, 1894, reads, "Mayhaws are ripening and the teeth of the small boy will soon ware a wire edge, but he will cut the mayhaws all the same."

Currently, during years of abundant crops, hundreds of thousands of pounds of mayhaws are gathered from Louisiana swamps by individual connoisseurs. A commercial market has also been developed, and it's now possible to enjoy a fine local mayhaw wine with the exquisite jelly on that buttermilk biscuit. Amen.

Bottomland Oaks

Oaks are dominant tree species in bottomland hardwood forests. They can be divided into two major groups: red oaks and white oaks. The white oak group has leaves with rounded edges and acorns that mature in a single season. Examples of white oaks that grow in lowlands include overcup oak, delta post oak, and cow oak. Leaves of red oaks are usually bristle-tipped, and their acorns take two years to mature. Lowland species include willow oak, water oak, cherrybark oak, and nuttall oak. Willow oak and water oak are commonly called pin oaks in Louisiana, but true pin oaks don't occur in the state and generally grow north of central Arkansas.

In natural settings oak trees are site specific by species. Lowland oaks are tied closely to land elevation, hydrology, and soil type. Overcup oaks grow on the lowest, wettest sites. Moving upward in elevation in the swamp, perhaps only a few inches, the oak component will change to willow oak and nuttall oak. These in turn will be replaced by water oak and cherrybark oak as one progresses upward to the subtle ridges along bayous and rivers.

Over thousands of years each species has adapted to grow best in specific site conditions. Changes in these conditions, either natural or manmade, can prove devastating to a species. A good example can be found along Bayou D'Arbonne in Ouachita and Union parishes, where hundreds of acres of willow oak are dying. The die-off is tied to the development of the Ouachita River Navigation Project, which raised water tables in the affected area. The trees, being unable to adapt to the new wetter site, are severely stressed, making them more susceptible to natural diseases and insects.

More than 7 million acres of bottomland hardwood forests have been cleared and converted to agriculture in the lower Mississippi Valley in the last century. Oaks were a primary component of this ecosystem and anchored the rich biodiversity. Every animal, from bears to songbirds, was tied directly or indirectly to oaks.

In recent years government conservation agencies have begun efforts to reforest some of the cleared areas that proved to be marginal farmland. Several thousand acres have been replanted with oaks and other native trees. The science of reestablishing naturally functioning bottomland forests is barely beyond its infancy, but successes to date are encouraging and a sign that oak forests can be maintained and restored if public attitudes so demand.

Upland Oaks

As was the case in bottomland hardwoods, oaks were once a common component of upland forests in north Louisiana. One native upland habitat type is classified as oak-hickory-shortleaf pine forest. Natural forests in the uplands are uncommon, most having been converted to loblolly pine monoculture for commercial purposes. In fact, the loss of native upland forests is greater than that in bottomlands, as estimates are that less than 10 percent remain.

Oaks are at the top of the wildlife food list in upland forests also. More than a hundred kinds of birds and mammals use oaks for food. Acorns, the staff of life for many species, are most important in the critical winter season when other foods are scarce. In all cases, removing oaks and other hardwoods from an upland forest will decrease biodiversity further by eliminating the animals dependent upon them. Such impacts flow downhill to affect wetland wildlife that once sought refuge in botanically diverse upland areas during natural flooding cycles of bayous and rivers.

Only a small fraction of Louisiana uplands are public lands that can be managed for natural biodiversity. The key lies with private and corporate landowners whose management decisions are usually market driven. Already, shortages of oak timber and problems with pine monoculture have convinced some to rethink old policies. A sustainable, multiple-use approach to managing the remaining native upland forests is critical in order to perpetuate oaks and their attached web of life.

Acorns

The white oaks of Union Parish released their offspring on November 19 of a recent year. Actually, they began a few days before and continued for a week or so afterward. On this day from my front porch my watch could not mark ten consecutive seconds free of acorn-fall within earshot. It was the largest acorn crop in fourteen years for this naturally cyclic species. As they fell at speeds up to 100 miles per hour they riddled the leaves below them like arboreal hail and buried their butts in Mother Earth. Some blasted the metal roof of my house, accelerated on the 7/12 pitch, and launched off the edge of the front porch at a very un-botanical angle. Once on the ground their troubles just began. Acorn-borers, fungi and other pathogens, birds, and mammals attacked this nutritional cornucopia with relish. Few survived.

Unlike their spring-germinating red oak cousins, white oaks germinate in autumn. A myth involving squirrels is entwined with this adaptation. Squirrels are commonly believed to assist in the planting of acorns as they bury them for winter food caches. Sometimes it happens, but gray squirrels are known to cut out the embryo of white oak acorns before they bury them. This keeps the acorn from germinating, which would result in a loss of food energy for the squirrel. Remarkably, squirrels do not excise the embryos of spring-germinating red oak acorns.

I counted 25 acorns in a measured square foot in my front yard. Eight of them, by the way, were infested with acorn-borers or were otherwise bad—better than normal. I also counted 44 white oak trees within 200 feet of my front porch. Each had an average crown of 1,300 square feet. Using my No. 2 lead calculator, the one with the pink eraser, I computed 1,430,000 acorns on the ground under the 44 trees. The 25 acorns that I collected weighed 3 ounces. All of the acorns in my yard thus weighed 10,725 pounds, and probably a third of the crop was still on the trees.

In spite of this tremendous reproductive effort by nature, the probability of even one of these acorns growing into a mature tree was almost zero. The old trees in my yard produce deep shade, and oak seedlings are shade intolerant. This means that unless I cut down enough trees to allow sunlight to reach the forest floor, the seedlings will never grow taller than a few inches. In natural conditions that rarely exist today, oak forests reproduced sporadically in tree-fall gaps. When an old, large tree fell, its offspring grew up in the sunlit gap. The aesthetics of a giant fallen tree in the front yard is of less mind to Mother Nature than to me.

Pines

Also in those areas a few rungs up the ladder from the lowest swamps, pines drop their needles into wintertime bayou waters. A group of about ninety species worldwide, most are found on drier, acid soils in the Northern Hemisphere. Some thrive only when fire is a recurring event in their lifecycle. Members of the pine family have no true flowers or fruit. Pollen and seeds are borne on the same plant in separate cones. Male cones are small, numerous, and clustered. Female cones are large and woody and contain winged seeds that mature in two years.

Five species occur naturally in Louisiana. Longleaf and shortleaf pines were historically the pines of the higher upland sites. Longleaf pine, named for its needles up to 15 inches long and also called pitch pine, is the most resinous of eastern pines, a trait that made it

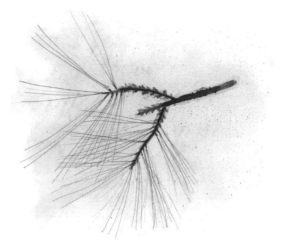

Loblolly Pine

America's leading source of naval stores for two hundred years. The "cat-face" scars of turpentine-gatherers were once discernable in the old longleafs of Kisatchie National Forest. Loblolly pines were the pines of damp, swamp borders, and slash pines grew in the wet flatwoods. Spruce pine, with its tightly furrowed gray bark and small cones, grew commonly in the rich hardwood forests of the Florida parishes.

Commercially, southern pines are among the most valuable of all trees. They provide lumber, millwork, plywood, and veneer for building construction and furniture. Chemical derivatives from pines include rayon, cellophane, many plastics, and some pharmaceuticals. Millions of tons of pine trees are processed into pulp for papermaking annually.

As a result of human manipulation loblolly is now the most common pine species in the state. Relative to other native pines it grows faster and in a wider variety of habitats. This species is the tree of most pine plantations. Genetically engineered to produce commercially desirable traits, loblolly has been planted on millions of acres throughout the South. Pine-based commercial forestry is a vital component of Louisiana's economy, but it has drastically changed the biotic structure of the historical ecosystem that overlaid the area. There was once 4 to 6 million acres of upland forests composed of a mix of oak, hickory, and shortleaf pine. Today only 5 to 10 percent of this forest type exists, as most has been converted to loblolly pine monoculture. Likewise, longleaf forests once blanketed 4 million acres in Louisiana; today less than 10 percent remains. A wide variety of plant and animal life thrived in the bountiful biodiversity of the historical natural forests. With the conversion the richness and abundance of life forms, from lady's slipper orchids to red-headed woodpeckers, plummeted.

Much has been written about how cotton, soybeans, and other crops have changed our natural landscape while fueling the economy. This may be true in the fertile riverine bottomlands, but pines in monoculture settings fill the niche of the two-bladed sword in the uplands.

Pecan

To me the word "PEE-can" is synonymous with the chamber pots of days past. However, a national survey conducted in 2003 finds that "PEE-can" over "pa-KAWN" is the overwhelming choice among Americans. I'll not conform to the majority.

The name "pecan" is actually of Native American origin and was used to describe nuts that required a stone to crack. Pecans are in the hickory family and grow naturally along the river bottoms of eastern North America and south into Mexico. Old, wild trees can exceed 100 feet in height and 3 feet in diameter. The well-known fruit of pecan trees was an important food for humans and wildlife for thousands of years before the first Europeans clanked ashore.

Native pecans exhibit great variety in nut size, shape, thickness of shell, and ripening date. Within this diversity an occasional highly desirable, wild tree was discovered with unusually large, thin-shelled, sweet nuts. In 1846, a Louisiana slave named Antoine successfully grafted one of these superior wild pecans onto a typical stock. His clones went on to be honored at the Philadelphia Centennial Exposition of 1876 and became the first official plantings of improved pecans. The successful use of grafting techniques led to grafted orchards and the widespread commercialization of pecan production. Today, there are more than 1,000 varieties of pecans with more than 300 million pounds produced annually in the United States.

The year 1704 was a rough time for Jean-Baptiste Le Moyne de Bienville's young Louisiana colony in New Orleans. Overdue supply ships from France resulted in a food shortage, and to stretch the remaining provisions Bienville released many of his men to go into the woods and live among the Indians until relief arrived. Andre Penicaut, a master carpenter, was one who left and traveled upriver to stay with the Natchez tribe. There, to his delight, he was introduced to a nut, which he described as "scarcely bigger than one's thumb." According to his spelling the Indians pronounced it "pa-KAWN."

Catalpa

Many of the younger folks seem to have drifted away from the once common outdoor life and the inadvertent absorption of knowledge al fresco. However, most every adult in bayou country old enough to remember Elvis in his prime can recognize a catalpa tree. We also call them "catawba" trees, substituting the "l" and "p" with a "w" and "b" in the way that we southerners slow down an uppity three-syllable word. Catalpas are deciduous, hardwood trees with heart-shaped leaves and long panicles of maroon and white flowers. They were formerly valued as ornamental shade trees and for durable lumber. Many were planted to line city streets across the state.

One group of admirers appreciates catalpas for reasons other than aesthetic. In fact, they enjoy them most when infested with a parasite and nearly defoliated. Resourceful fishermen have long sought the parasites in the form of caterpillars as fish bait supreme. Indeed, many a whiskered catfish has been wrestled over the side of a cypress bateau after falling victim to a 2/0 hook garnished with a catalpa worm. We do call them worms, but they are actually caterpillars of the catalpa sphinx moth. They begin their lifecycle as eggs attached to the bottom of catalpa leaves in the spring. The eggs hatch into caterpillars that voraciously eat the leaves, often leaving the tree in a near-naked state. If the caterpillars manage to avoid predators, such as wasps, yellow-billed cuckoos, and fishermen, they enter the ground under the tree and pupate into adult catalpa sphinx moths. The plain, brown moths emerge and lay eggs on the tree to complete the cycle. Several generations occur in one summer, and the last one over-winters in the ground until the following spring.

So, opinions concerning catalpa trees are not unlike politics—a matter of perspective. One either votes for them as handsome ornamentals or as sources of fish bait. You can't have your shade tree and eat your blackened catfish, too.

The adage of not being able to see the forest for the trees is ripe with biological truths, as most plant species in a natural forest never approach the canopy in stature.

May-Apple

When my dad was a boy tromping about the red clay hills that hemmed in Bayou D'Arbonne, an odd-looking herbaceous plant always foretold the coming of spring. May-apple, sometimes called mandrake, poked its umbrella-shaped leaves up early to capture vital sunlight under the naked hardwoods before being sentenced to shade for life. Once common in moist, fertile soils throughout the eastern United States, may-apple has declined in the South under the advance of diversity-squelching pine plantations. The remaining plants are found in clumps and grow to about 18 inches tall. A single white flower blooms locally in April and later forms a crabapple-size fruit. By late summer, all above-ground evidence of this delicate perennial has vanished in the subtropical heat.

Native Americans were the first to recognize the practical values of may-apple. The fully ripe fruit is edible and was once made into jams, jellies, and pies. All other parts of the plant are poisonous to some degree. Indians used the root to treat internal parasites and as a strong laxative. The cathartic properties were valued later when it became a component in Carter's Little Liver Pills. Today may-apple and a closely related Asian species are best known in the medical field for a chemical found in the roots called podophyllotoxin. This very strong plant alkaloid is thought to protect may-apple from insects and other herbivores. Acutely toxic, it is now an active ingredient in a drug used to treat lung cancer. During chemotherapy it inhibits the activity of an enzyme necessary for cancer cells to replicate.

May-Apple

May-apple still grows in my heavily forested yard just up the hill from Bayou D'Arbonne, but I see it in a different light. On the morning that I write this essay it flows into my father's ailing body through an IV port. It is his chance to enjoy the coming spring once again.

Sunflowers

It may come as a shock to many people to know that sun-worshiping pagans are common in every parish of Louisiana. Not unlike the ancient Greeks, Egyptians, and Inca, they track our star across the diurnal sky with faces uplifted in joyful adoration. They are actually members of one of the largest families in the world—a family of plants, that is. The sunflower family contains nearly 24,000 species, about 1 of every 4 kinds of plants. They grow worldwide but are most common in tropical and temperate regions. There are 396 varieties known to live in Louisiana, from the red clay hills down to the marshlands. Family members are diverse and consist of trees, shrubs, vines, annuals, and perennials. All are characterized by unusual flowers. What appears to most observers as a flower is actually an inflorescence that consists of many small individual ray or disc flowers, or a combination of both.

Familiar sunflowers include the type used for birdseed and those commonly found in gardens, such as zinnias, marigolds, daisies, and cosmos. Dandelions, thistles, ragweed, goldenrod, sagebrush, saltbush, and cockleburs are also sunflowers. A healthy salad contains sunflowers in the form of fresh lettuce, and we consume parts of other sunflowers like artichokes, chicory, safflower, chamomile, and echinacea. Extracts of marigolds are used in the poultry feed, cigarette, and cola industries. Others are considered medicinals and important honey plants. Except perhaps in the dead of winter, it is difficult to experience a Louisiana day without being exposed to some form of sunflower.

Poison Ivy

At the same time President Thomas Jefferson sent Meriwether Lewis
and William Clark to explore the Missouri and Columbia rivers,
he commissioned William Dunbar to conduct a similar expedition
on the Ouachita River from its mouth to the legendary Hot Springs
in present-day Arkansas. Dunbar's mandate was similar to that of
Lewis in that he was required to record and describe native plants
and wildlife observed during the journey. One passage in his journal reads, "We have a Vine called the poison vine, from a property
it possesses of affecting some persons passing near it, by causing an
inflammation of the face resembling an Erysipelas. Other persons
may handle this vine with impunity. It is believed perhaps without
reason, that some are affected by only looking at it."

Fortunately, you can't get poison ivy by just looking at it, although for some folks it doesn't seem to take much more than that.
Poison ivy is one of three plants in bayou country that cause a contact dermatitis in many people. It is a vine that grows to 50 feet or
more and has characteristic three-lobed leaves. It is found in all
habitat types. Much less common are poison oak, a small shrub that
grows only in dry upland soils, and poison sumac, a small tree with
divided leaves containing up to eleven leaflets. All are in the cashew
family.

These plants contain the poisonous oil urushiol. When it comes
in contact with skin, the chemicals cause an immune reaction, producing redness, itching, and blistering. It is important to remember
that you don't have to touch the plant to have a reaction. The oil can
be carried on the fur of pets, on garden tools, or on any object that
has come in contact with the plant. It can even be transmitted in
the smoke of burning poison ivy vines. Your momma was wrong on
one alleged method of conveyance though, as it can't be spread by
scratching the blisters.

Sensitivity to poison ivy is not something we are born with. It
develops only after several encounters with the plant. Studies have
shown that approximately 85 percent of the population will de-

velop an allergic reaction if exposed. Sensitivity varies from person to person. Although they are not sure why, scientists believe that an individual's sensitivity changes with time and tends to decline with age. Those who were once allergic may lose their sensitivity later in life.

Fruits of these poisonous plants are consumed by many kinds of wildlife without any apparent ill effects. Deer relish poison ivy leaves and concentrate the toxin in their chambered stomachs, creating an occupational hazard for careless biologists who sometimes must examine them during herd health checks. This I can vouch for personally.

As usual, prevention is the best cure. Know how to recognize the poisonous plants and avoid them. If you are exposed, wash with cold running water as soon as possible. See a doctor for intense cases—and don't worry about catching poison ivy just by looking at it.

Orchids

Is there a southern twayblade or fragrant ladies' tresses in your life? Perhaps not, but they and their kin occur in numbers and an array of diversity that might surprise you. Orchids are often thought of as exotic, gaudy, almost unnatural flowers found only in jungles and corsages. Actually, they make up the largest family of flowering plants in the world, with more than thirty thousand species. Several are very common in Louisiana.

Orchids are unique in many ways. Most species found in the tropics are epiphytic—growing above ground attached to tree bark and deriving their moisture and nutrients from rain and air. However, with one exception those that live in Louisiana are terrestrial, growing in the ground. Orchids are considered the most specialized of flowering plants and will only grow in habitats with very specific conditions. They produce the smallest seeds of any flowering plant, and one plant may release more than a million of the dust-like particles. Once the seeds germinate, growth will not occur without the presence of mycorrhizae, a special type of soil fungus. The fungi actually penetrate the cell of the seeds and provide nutrients for the growing plant. Development of the mycorrhizae/orchid relationship is slow, and in some species it takes ten years for the plant to appear above ground and flower.

Native orchids in Louisiana include the crane-fly orchid, common in the hill parishes. For most of the year it lives as a single purple-bottomed leaf on dry upland hardwood/pine sites. The water-spider orchid that grows more than 2 feet tall lives at the other end of the hydrologic spectrum. It is found in masses of floating aquatic vegetation in swamps and bayous throughout the state. Because each native orchid species has very different and exacting needs, they should not be removed from the wild. Precise amounts of sunlight, nutrients, moisture, and the presence of critical mycorrhizae are extremely difficult to mimic in a cultivated setting. Native orchids grow where they grow for a reason, and, unlike many cultivated plants, the reasons have nothing to do with the desires of humans.

Adder's Tongue Ferns

Beginning in the early 1970s, strange activities started to occur during early spring in graveyards throughout Louisiana. Reports indicated bizarre behavior by small groups of people in cemeteries both rural and urban. To observers these people were obviously not there to pay respect to deceased loved ones or friends, as is usually the case with visitors. They were dressed in rugged field clothes; some were shabby in appearance. Most of them were young, but there was always an older, balding man in their presence, obviously the leader of their rituals. The scenario was the same at each event. The group would arrive at the graveyard in a hodgepodge of vehicles and immediately gather around the leader for a blessing of sorts. With plastic bags in tow they would then disperse across the cemetery to begin the really weird goings-on. As soon as each individual reached a mysteriously chosen spot, he or she immediately fell to his or her knees and began crawling slowly about on all fours with butts often higher than noses. It seemed to be a quest for some tiny, ghoulish treasure, and when the object was found the discoverer emitted a screech of pleasure that straight away caused a mini-stampede as everyone rushed over to worship the object. This went on for a while with varying degrees of enthusiasm or despondency depending on how many totems were found, until the entire group loaded up and drove away as inexplicably as they had arrived. The activity continued for several successive springs—or so it all seemed.

The truth of this matter when anyone bothered to ask was about as strange as the speculation. The leader of the group was a prominent professor of botany at Northeast Louisiana University and the others were his students of plant taxonomy. They were searching for botanical treasures in the form of any of five species of a tiny plant called adder's tongue fern. Very un-fern-like, the plants consist of a single, simple, ground-hugging leaf less than 1 inch long. One has to

Adder's Tongue Fern

be purposefully and intently looking to find it. It is most common and most easily found in areas that have been mowed closely for many years, thus the cemetery searches. In their diligent hunts the professor and his students contributed to science by expanding the known range of these little-known plants—facts usually lost on curious passers-by.

Vines

As a child I was amazed by the Tarzan movies. I could never figure how the Lord of the Jungle always had a convenient vine from which to swing out of danger or into the arms of Jane. Also in defiance of all probabilities, when Tarzan reached the apex of a swing only to alight gracefully on a massive limb, there was always another vine ready to carry him on his arboreal way.

A vine can be broadly defined as any climbing plant. Vines' ability to climb is an adaptive feature that allows them to monopolize sunlight by using other plants as support structures. By climbing or creeping along the ground, vines can also keep their roots in favorable soil and grow out over poor soil or rock surfaces where other plants can't grow in order to reach more sunlight with little competition.

Vines can be found in several different plant families using different techniques to climb. Only a few, like morning glories, resemble the typical artists' renditions of plants twining their stems around a support. Some, like poison ivy, have clinging roots to work their way up a tree. Others have tendrils that can be specialized shoots, as in muscadines and other grapes, or even modified flowers, as in passion flowers. Virginia creeper has twining stems with adhesive pads that firmly attach the vines to a support.

Vines are a natural part of many ecosystems. Various species of birds and other animals use vines as a food source—think grapes and greenbriar and rattan berries—and places to build their nests. Some vines can live to a very old age, and biologists use the presence or absence of large, old vines as an indicator of the ecological health of a tract of forest. With the exception of deep overflow swamps, most forests in Louisiana once contained many large vines. Their scarcity now is a result of intensive logging.

Unfortunately, when non-native vines are introduced into a new area they can become an invasive species to the detriment of native flora and fauna. Japanese honeysuckle and Chinese wisteria are now with us for the long term, and it just occurred to me that the Tarzan vines might really have been kudzu.

Lichens

The next time you get a chance, take a close look at a tree trunk. It's pretty likely you'll discover gray-green patches of an organism called lichen clinging to the bark. Now, take a deep breath—lichens are indicators of clean air.

Lichens are a good example of nature's creativity. They represent the cohabitation of two organisms, fungi and algae, from totally different taxonomic kingdoms. A fungus does not contain the photosynthetic pigment chlorophyll and thus cannot make food out of sunlight. The fungus, though, provides structure for support of the organism and helps prevent drying out. The alga is a photosynthetic plant and can make organic nutrients from sunlight, so it provides the food. This ability of organisms to live and work together is called symbiosis.

Louisiana forests and swamps are rich in lichens. They are classified by three basic shapes: crusty lichens lay flat on tree branches or rocks, leafy lichens are the most common and have a leaf-like form, and fruticose lichens are erect and branching. Examining the wondrous forms of lichens with a 5X hand lens can provide hours of pleasure for kids and kids at heart. One writer compared the experience to snorkeling over coral reefs in the ocean.

Lichens have both natural and artificial uses. The ruby-throated hummingbird builds a tiny, cup-shaped nest out of lichens, spider silk, and fern fuzz and attaches it to the top of a branch. The "reindeer moss" of the Arctic tundra is actually lichen that is a staple in the diet of reindeer and caribou. In Louisiana, white-tailed deer eat lichens, although not as a preferred food. A few years ago on the Hebridean island of Harris in the North Atlantic I was exposed to a centuries-old use of lichen. The island is famous for its wool fabric known as Harris Tweed. The preferred agent to dye this wool came from a gray lichen called crotal, which grows slowly on rocks and ancient stone walls. Women and children used soup spoons to scrape the lichens from the rocks. Lichen is also used to make litmus paper.

Air pollution and lichens are not compatible. A recent visitor to the Louisiana bayous from Miami, where lichens have disappeared, was enthralled with the miniature world of lichens here and referred to them as "inconspicuous wonders." It might behoove Louisianans to at least keep an eye on them.

Mistletoe

Well, the Druids thought it peculiar too. If you are traveling around Louisiana, scan the tops of hardwood trees and look for the dark green clumps of mistletoe. How did they end up growing in the loftiest boughs of our tallest oaks?

More than twenty species of mistletoe live in North America with others in Europe. The most common type in Louisiana has fragile, green stems and small, opposite leaves. Clumps of white berries form in late autumn. Eastern mistletoes grow on broad-leaved hardwood trees, while most of those found in the western mountains and the Pacific region grow on evergreen conifers, such as pine and spruce. Mistletoe is parasitic on its host tree, deriving most of its water and nutrients in the form of minerals from the branches to which it is attached. Although a heavy growth of mistletoe may contribute to the decline of a tree with other ailments, it doesn't usually kill its host.

To the Druids, mistletoe appeared to spring from thin air.

Equally strange, it seemed to defy nature by living its entire life high in the branches of trees, never descending to earth, a plant's natural habitat. For these reasons they declared mistletoe and the oak trees on which it grew sacred. Six days after the new moon, white-robed priests gathered mistletoe with a golden sickle, and, following prayers and the sacrifice of two white bulls, brewed a potion with special health-giving properties. Or so they say.

For centuries after the demise of the Druids, mistletoe continued to show up in medicinal potions. In Sweden, mistletoe was a key ingredient in a concoction that also contained live baby swallows and was promoted as a remedy for sore throats. This "swallow's water," as it was known, remained in favor until 1757, when the Swedish government mercifully banned it. According to Louisiana folklore, early settlers used mistletoe berries to treat people who had fits. The dose was a thimbleful of ground berries four times a day. The powder was also given to dogs suffering from fits. Its reputed sedative properties may account for this use. At various times it was also said to cure tuberculosis and stroke. Modern research has found no conclusive medical benefits of mistletoe, and it is usually considered poisonous. The U.S. Food and Drug Administration lists this plant as "unsafe," and it should never be taken internally.

How *does* mistletoe become established in treetops? Birds of course are the culprits; after eating and digesting the berries, they scatter the seeds on the next convenient perch. In Louisiana the berries are relished and thus dispersed by bluebirds, robins, and cedar waxwings.

Mistletoe does not, as some once believed, descend in a flash of lightning from the sky to alight on the sacred oak. My advice concerning the legends of mistletoe is to heed only the one that encourages holiday kisses.

Prairies

The term "prairie" usually brings forth images of treeless landscapes blanketed in a waving sea of grasses reaching to the horizon.

Indeed, such was once the case across millions of acres in the American West and Midwest. This unique habitat nourished bison herds that numbered in the tens of millions together with countless other species of plants and animals.

Prairies are often classified as either short-grass or tall-grass prairies, depending on the dominant types of grasses that grow there. Short-grass prairies were generally found farther west in areas of low rainfall. Certain soil types often favor the development of prairies, but many are under constant attack from invading woody plants. Just as an abandoned lawn in Louisiana will eventually revert to brush and then trees, prairies will eventually turn into savannahs or forests in many cases without some type of intervention. Intervention came naturally in the form of huge herds of bison eating and trampling everything in their path during cyclic migrations. It also occurred as fire from lightning strikes or intentionally set by Native Americans. In any case prairie plants thrived with disturbance, and the invading woody species were set back.

In Louisiana, tall-grass prairies once covered more than 100,000 acres in the southwest coastal area. Less than 1 percent remains, and most is found on old railroad rights-of-way. In north Louisiana prairies were much less common and smaller in size. In 1783, Don Juan Filhiol, Spanish commandant of the District of Ouachita, made his first permanent settlement at a place called Prairie des Canots where Monroe now stands. In 1812, Amos Stoddard, when describing the same area, wrote, "on the left bank of that river are extensive prairies, the soil of which is luxuriant and productive, bearing a high coarse grass." The vil-

Purple Coneflower

lage of Oak Ridge in Morehouse Parish was once known as Prairie Jefferson, and early maps note Mer Rouge as Prairie Mer Rouge.

Today only a few remnant areas of native prairie are known in Louisiana. Efforts to restore prairies on a small scale are promising, and in addition to the grasses, spring brings forth a profusion of purple coneflowers, Indian blankets, and wine cups to remind us of what was once not so rare.

One-fourth of the plant species growing in bayou country are not native to the state. Some are beneficial neighbors, some innocuous, but others can diminish our world.

Chinese Tallow

Okay—this is not going to be pleasant at first, but you've got to do it. Go to the garage, get the axe, and chop down that Chinese tallow tree in your yard. For good measure dose the stump with Roundup. Now, don't you feel better? Well, you should, because you just eliminated a botanical terrorist no less threatening than kudzu or water hyacinth.

Throughout Louisiana, Chinese tallow tree is among the worst of our invasive species. This native of Asia has become established throughout the southern coastal plain from North Carolina to Texas. It is a small to medium-sized tree up to 40 feet tall with heart-shaped leaves that are simple, alternate, and deciduous. The seeds are white and have a waxy, vegetable tallow coating. After the first autumn frost the tree plays its siren song in a brilliant splash of autumn color. It is a lure to be avoided at all cost.

The problem with this displaced species is that it rapidly replaces native plants. It is presently doing almost irreparable harm to our wetlands and other natural areas by reducing biodiversity. Fallen leaves of tallow tree alter soil chemistry and are poisonous to other plants. Entire ecosystems can be disrupted. Imagine the cypresses and oaks along your favorite bayou replaced by thickets of scraggly tallow trees. It has already happened in many places.

Moving water and birds disperse tallow tree seeds. When the tree in your yard produces the first seed, it is forevermore guilty of contributing to the degradation of our natural areas. If you seek autumn color, there are alternatives. First, go to the natives—red maple, sourwood, even sweetgum can rival the foliage in Pandora's box. If you are domesticated beyond rehabilitation, then plant a Bradford pear. Though an alien also, it has beautiful color but lacks the wanderlust of a tallow tree. So, don't be known as one who harbors terrorists. Sharpen that axe and save the world. It will feel good!

Chinaberry

The wood of mahogany trees yields some of the most valuable lumber in the world. Cherished for its beautiful luster and resistance to rot, mahogany is native to Asia but grows unappreciated in Louisiana. It is found in the yards of old house places, especially those of tenant farmers in the delta lands. The Chinaberry tree is a type of mahogany with wood just as attractive as that found in fine furniture. Introduced to North America in the late eighteenth century and having no natural enemies, it has colonized the southern half of the United States and the eastern seaboard. The tree may have been planted around house places because it grows fast and provided shade in a sea of cotton fields. Rural people used the yellow marble-size fruit of Chinaberry to make whiskey and soap. The bark of the root was used to treat intestinal parasites, and the leaves were said to discourage bott fly larvae in horses. Leaf litter from Chinaberry causes the soil to become more alkaline and discourages other plants from growing nearby. The actual benefits of Chinaberry products and homegrown remedies may have been more harmful than helpful because the berries and other parts of the plant are moderately toxic to humans and livestock. Even today Chinaberry poisoning occurs, with pigs and dogs most often reported as victims. Birds like robins, however, seem immune to the poison and only exhibit varying degrees of intoxication after overindulging in the fruit. Modern medical research indicates that a product in Chinaberry leaf tissue may be effective in treating the human herpes virus, something the delta sharecroppers could not have imagined when they planted the trees as a respite from the southern summer heat a hundred years ago.

Osage-Orange

On Thursday, May 7, 1857, a plantation diary reveals that a slave named Hastings was put to work trimming a dense, thorny hedge around a field adjacent Bayou Bartholomew near Bastrop. Six years later, on June 7, 1863, Major General J. G. Walker of the Confederate Army attacked a Union force at Milliken's Bend in Madison Parish in hopes of relieving pressure on the besieged fortress of Vicksburg. His attack was thwarted, in part, because of a dense hedge around part of the village. Yankees massed behind the hedge and fired through the openings. General Walker wrote, "Upon reaching the hedges it was utterly impracticable to pass them except through the few openings left for convenience by the planter. In doing this, the order of battle was necessarily broken."

The plant that contributed to the slave's misery and the Rebels' frustration was osage-orange, also known as bois d'arc or horse apple. Originally, it grew along the Red River Valley in Texas and Oklahoma. The name of the tree comes from the Osage tribe, who lived in that area and valued the strong, elastic wood to make bows. The spread of the species into other areas began as the Osage traded it among Plains and southeastern Indian groups. White settlers quickly learned to use the tree to create impenetrable living fences before the invention of barbed wire. Saplings were pruned to promote a bushy growth. "Horse high, bull strong and hog tight" were the criteria for a good osage-orange hedge. This meant that it was tall enough that a horse could not jump it, strong enough that a bull could not push through it, and woven so tightly that not even a hog could root through it.

Osage-orange is in the mulberry family. It grows to 40 feet tall and is known for its unusual fruits, which are hard, warty, yellow-green, and about the size of a softball. Squirrels relish them. The wood is still used for fence posts, and at one time the bark was used for tanning leather and making a yellow dye.

The famous expedition of William Dunbar and George Hunter up the Ouachita River in 1804 made the first known scientific documentation of osage-orange in North America. Two hundred years ago, they recorded it as growing upstream of what is now Monroe, and in their report to Thomas Jefferson said that these plants had been transplanted from somewhere else. Today it is still possible to find descendents of the early hedges scattered as individuals along the high banks of the Ouachita and Red rivers and bayous such as Macon, DeSiard, and Teche. They harbor a bit of local history under their gray, ridged bark.

Only in the Bayou State can extinct flora generate legislative action.

State Fossil

In 1976, while most of the nation was celebrating America's bi-centennial, the Louisiana state legislature was up to more important things. They were debating the designation of an official state fossil. Apparently, the issue became contentious when one senator nominated a colleague for the title. Calm returned to the chamber floor only when the second senator declined in deference to age rather than beauty. Subsequently, the distinguished body voted unanimously to name petrified palm wood as the official state fossil.

Fossils in Louisiana are relatively scarce, and petrified wood was a good choice. Petrified wood is formed when any of several types of minerals replace buried woody tissue. Silica is the most common replacement mineral. In the western United States much of the petrified wood developed after being buried by volcanic activity. In Louisiana the wood was buried in the silts and sands of meandering rivers and streams that occurred on the Gulf coastal plain around 30 million years ago. The shore of the Gulf of Mexico was farther north then, explaining why most petrified wood is found in the northern half of the state.

The Louisiana state fossil is specifically petrified palm wood. Of the many types of petrified palms, those found in Louisiana are most commonly in the genus *Palmoxylon*. It is a favorite of rock collectors because of high silica content, well-defined, rod-like structures, and variety of colors. Jewelers like it because it polishes well and is durable. They follow in the tradition of Native Americans, who used worked petrified wood as tools for thousands of years.

While some might argue that Louisiana politicians are indeed petrified in their governing abilities, their efforts to recognize an interesting fossil should be considered on their educational merits.

2
BAYOU-FAUNA
Animals in a Place of Bayous

Bear chewed up permit station on Mill Rd. at 10-Lick line; [I] caught a
3' canebrake rattler & released him

    —KO Field Diary, 10 June 1986—Tensas River NWR

Law Enforcement patrol Mollicy early—saw 3 otters, bald eagle, buck,
coyotes, lots of raptors & flight ducks

    —KO Field Diary, 21 October 2003

More than 450 species of wild birds have been recorded in Louisiana. From bayous in wetlands to hilltops in uplands, birds blanket
our landscape, settling into niches as varied as a barren spit of sand
to an old, fungus-infested pine. Some are static, completing their
lifecycle in back yards, but most participate in the rhythmic arca-
num of migration, a behavior that mocks rational thought. Even before John James Audubon painted bayou birds to fame they were the
ornaments of Louisiana fauna. Birds in their infinite diversity continue to embellish our imaginations.

    About 150 species of freshwater fishes swim the bayous, rivers,
streams, lakes, ponds, sloughs, oxbows, and marshes of Louisiana.
They are as diverse as their habitats. A few are familiar as gamefish,
and others were once components of a thriving commercial fishing
industry. Most, though, are unknown to the average citizen and yet

fill important niches in the web of life just below the surface of our consciousness.

The term "herps" is a colloquial derivative of "herpetofauna," which refers to the group of animals consisting of amphibians and reptiles. Louisiana bayous and swamps often evoke images, usually negative, of an abundance of these creatures. In fact, about 130 species of herps are found in Louisiana, including 27 frogs, 22 salamanders, 14 lizards, 27 turtles, 39 snakes, and 1 crocodilian. The numbers vary slightly according to which herpetologist does the counting. Herps are ubiquitous in Louisiana although some species have very specialized niches. Generally misunderstood and often persecuted, herps play important roles in the natural world.

Louisiana wild mammals run the gamut of diversity and include bears, bats, deer, dolphins, and many creatures in between that nurse their young. Some species are well known with economic and recreational value (e.g., deer, furbearers); others are virtually unknown by most people (e.g., bats, shrews). All are integral components of the various ecosystems across the state and often involved in the checks and balances, vital interspecific links and strands, which maintain the natural world.

They are spineless. They have absolutely no backbone. Many are blood-sucking leaches and parasites of the worst kind. Some have an uncanny ability to make our lives miserable, but we need them. No, I am not referring to shady politicians here but rather those creatures known as invertebrates. Invertebrates are all the types of animals that lack backbones. They are by far the most numerous animals on Earth. Two million species have been identified, which make up 98 percent of all known animals. They inhabit every habitat on the planet, from tropical forests to alpine tundra, from the depths of caves to seabed mud. They drift the ocean currents as plankton and ride desert winds as the aerial dust of life. Most invertebrates are small, some as tiny as bacteria. Others like the giant squid can be 60 feet long and weigh 4,000 pounds. One type of ribbon worm has the diameter of a pencil and is 180 feet long.

Invertebrates are divided into several groups, of which arthropods is the largest. It consists of insects, like flies, cicadas, moths, earwigs, fleas, cockroaches, bees, beetles, dragonflies, and termites; crustaceans, such as lobsters, crawfish, shrimp, crabs, barnacles, and pill bugs; and spiders and their kin. Another important assemblage is the mollusks, which include clams, mussels, snails, seaslugs, octopus, and squid. Centipedes and millipedes are thrown into the melee for good measure. Other major groups include sponges, jellyfish, echinoderms like starfish, and several groups of worms. Some invertebrates have less than pleasant relationships with people, at least from the human perspective. A few cause parasitic diseases in humans and domestic animals. Others are agricultural pests, destroying plant crops. Wasps, mosquitoes, ticks, and jellyfish are just aggravating to most folks. But we can't live without them. Invertebrates occupy several tiers on the food web, including consumers, producers, and decomposers. Earthworms fertilize our soils, and insects pollinate plants of the planet. Invertebrates are used in medical research, in the development of drugs, and as monitors of the environment. And what about shrimp etouffee simmering in the old kitchens along Bayou Teche, Oysters Rockefeller at Antoine's in New Orleans, and that migrating monarch butterfly drifting across an azure autumn sky?

White-footed Mouse

A houseguest of ours left recently. Now, I love company. We have a big house in the woods with lots of room for folks to spread out if necessary—but three months is a bit long, even for the best of friends. He didn't say a word when he left, either. I have to admit that I had been encouraging his departure for quite some time even to the point of—dare I say it—attempting his assassination on more than one occasion. He drove me to it, you see. It started when he began eating the dried flower arrangements in the living room. This bit of thoughtlessness was only the prelude to fractious rude behavior.

White-footed Mouse

This guest was a mouse, a native white-footed mouse, sometimes called a deer mouse. He was not to be confused with the common, smelly house mouse with the very fitting scientific name of *Mus musculus.* House mice immigrated to Louisiana along with your ancestors from the Old World. Deer mice are generally forest dwellers and were already here to greet them when they got off the boat.

Deer mice are actually very handsome. They have white feet and stomachs, which contrast sharply with their dark backs. Large ears and huge black eyes make them easy to like when first introduced. One begins to question their character, though, upon learning their sexual mores. Females produce their first litter, usually four naked pups, at only 6 weeks of age. Breeding continues year round but for a marked lull in summer. It seems that the fertility of males declines significantly when the ambient temperature exceeds 89 degrees Fahrenheit. Perhaps that's why my guest left. He chose as a bedroom/dining room the space between two floor rafters just underneath the shower. As a result, he was exposed nightly to steaming saunas that very well could have left him impotent. Don't get me wrong. I like deer mice fine so long as they stay in the woods where they belong. Everything else likes them there too—as food. Owls, hawks, snakes, foxes, and other predators ensure that they rarely die of old age. The straw that broke the camel's back for me was when he rolled individually a dozen hickory nuts from a bowl in the living room 40 feet across a hardwood floor to his cache in the bathroom at 2:00 a.m. I baited the snap traps.

Purple Martin

A while back, during the month of July, email messages between birdwatchers from throughout the state were buzzing about a remarkable natural phenomenon occurring in downtown Shreveport. It seems that purple martins decided to establish the largest communal roost of that species on the entire continent in several trees around the Barnwell Center.

Purple martins are a type of swallow, the largest in North America. Adult males are uniformly blue-black, and the females have dark backs and gray bellies. Their call is a rich, gurgling warble. Nesting usually begins in April, when three to five eggs are laid in a nest of leaves and mud plaster. Incubation lasts for about two weeks and the young fledge a month later.

We usually think of martins as spring birds and eagerly look forward to the arrival of the first "scouts" in early February. They readily accept artificial nest boxes and seem to thrive in the close company of humans. Native Americans were fond of martins and erected gourds as nest boxes long before Europeans arrived. Purple martins originally nested in tree cavities that were abundant before modern short-rotation, industrial forest management. Competition from introduced house sparrows and starlings further reduced their natural habitat. Despite thousands of artificial nest boxes, martins still suffer from a housing shortage.

My brother and I decided to take a road trip to Shreveport to check out the electronic gossip. We arrived at the riverside Barnwell Center complex about 7:00 p.m. with nary a martin in sight. It was, however, quite obvious that something strange was going on. The nearby live oaks, shrubbery, plastic domed greenhouse, and sidewalks were covered with a white substance that we, as astute biologists, quickly determined not to be spray paint. Soon, lone martins appeared, circling lazily on the afternoon thermals. People also started gathering, walking from distant parking lots even though ample parking was available nearby. Some were carrying umbrellas on this cloudless day. It was, as the news people like to say, "a developing situation." Martins were becoming more common in every direction now as the sun slipped below the horizon. People began to leave the center of the park garden. Some stood under covered walkways. As the light faded, a sense of urgency enveloped the area. Martins by the hundreds, thousands, tens of thousands, began whirling in circles with ever decreasing radii around us. It seemed as if troops of winged tornados were converging on a few acres of sacred ground. One landed in the tallest cottonwood, and soon the branches would hold no more. In descending order of height the nearby trees were smothered. More feathers than leaves covered the limbs. For twenty more minutes they reeled and swirled closer and closer to the earth until all were settled in the darkness.

In a few days they were gone, most on the way to Brazil for the winter. People who do such things estimated their peak number at 1,200,000 on July 30. This unforgettable spectacle raises questions. Where did they come from? Why did they choose these particular few acres in northwest Louisiana as a staging area? Where have they been in previous years, and will they return in future years? On a more personal level, how did I, sans umbrella, escape unscathed?

Bryozoan Colony

Okay, here's the Louisiana bayou trivia question of the day. Just what are those softball-sized, jelly-like globs that are often seen attached to bayou trees and boat docks, especially after water levels fall? Impress your friends with this answer: bryozoan colony. These gray, gelatinous masses are actually colonies of thousands of individual animals called zooids. Each zooid is a microscopic creature complete with a mouth, digestive tract, muscles, and nerves. The jelly-like material serves as a protective matrix for the colony. Individuals feed by filtering tiny algae from the water through tentacles. Since algae don't usually grow well in muddy water, the presence of bryozoan colonies in a stream can be an indicator of good water quality, at least in terms of turbidity. Colonies grow in size by budding from the adult zooids. New colonies are established from free-swimming larvae produced by the zooids. There are many species of bryozoans, but most live in saltwater environments. Of the approximately twenty freshwater species found in North America, most live in warmer regions.

Studies have shown that humans are much more likely to develop an affinity for animals that have soft, furry coats and large eyes than for creatures without backbones. So where does that leave the blind, slimy, bayou-dwelling invertebrates that make up the bryozoan colony in terms of popularity? Well, as long as we don't pollute all of our waterways, it probably doesn't matter. Bryozoans have been around for 450 million years according to the fossil record and will probably be here long after we've stopped asking trivia questions.

American Eel

One of the most profound mysteries of nature involves salmon and their epic journeys from the ocean back to the freshwater rivers and streams of their birth in order to reproduce. In Louisiana there exists a species of fish that performs a feat no less amazing. In fact, this species does the salmon act backward. I am referring to the American eel.

Eels are widespread in North America and common in the rivers and larger bayous of Louisiana. Although their bodies are elongated and snake-like, they are actually scaleless fish with fins and gills. They should not be confused with what many people call "lamper eels," which are not eels at all but rather a species of harmless salamander that frequents swampy areas and ditches. True eels are sometimes called "fish eels" in this region. They occasionally reach

5 feet in length and are mostly nocturnal, feeding on a variety of fish, insects, snails, and crawfish.

There is much yet to be learned about the lifecycle of this species, but what is known is remarkable. Behaving exactly opposite of salmon, eels live out their adult lives in freshwater and return to the ocean to spawn. In fact, all eels return to a specific area known as the Sargasso Sea just north of the Bahamas. Here their life begins and ends. Eggs hatch into 2-inch larvae and drift in Gulf Stream currents for up to a year, most eventually arriving on the eastern coastline of North America. Some, though, are drawn into the Gulf of Mexico. As they drift they change into a more eel-like form, and usually in the autumn when they are still less than 4 inches long they begin to enter freshwater rivers and bayous. The young eels are determined at this stage and have been known to climb the wet walls of dams and wiggle up moist grass banks to get around obstacles. Many travel upstream several hundred miles where they may live as adults from 5 to 20 years. At some point the adults begin drastic physical changes that prepare them for migration back to the sea. They stop feeding, their eyes and fins enlarge, and their body color pattern transforms. The migration occurs during autumn nights as they retrace their natal routes down rivers and bayous, through locks and over dams, and back into the ocean for a January spawning in the warm Caribbean waters. Here the females lay 10 to 20 million eggs each, and life for the species is renewed even as the adults soon die on the spawning grounds. This profound mystery occurs right here in our midst and always below us as we cross the bayou bridges with minds on the mundane issues of our own lives.

Darters

The most popular kinds of freshwater fish in Louisiana are bass, crappie, sunfish, and catfish. They are well known because they are fun to catch and good to eat. However, in terms of biological diversity this group falls at the bottom rung of the aquatic ladder. In Louisiana there are only two species of black bass, two species of crappie, six species of catfish, and nine species of sunfish. Another group, almost completely unknown even to avid fishermen, swims our bayous and creeks with dramatic diversity of form, color, and species. Collectively they are called darters.

Darters are the second largest family of North American fishes; only minnows have more species. They are found only in North America and only in freshwater. Some are widespread, and others are restricted to single streams, leaving them vulnerable to extinction. Twenty species of darters have been found in Louisiana. Rarely longer than 3 inches, most inhabit clean streams with sand or gravel bottoms in the Ouachita and Pearl River drainages. Bedecked in vivid greens, blues, reds, and oranges, their spectacular colors never fail to amaze first-time viewers. That such gaudy creatures could exist in our midst almost unknown is always surprising. Even their names reflect the diversity: rainbow darter, harlequin darter, cypress darter, bluntnose darter, goldstripe darter, speckled darter, redfin darter, and banded darter.

Animals such as darters often elicit the disturbing question: "So what good are these critters anyway?" You are not likely to hear about the first invitational Caney Lake darter tournament any time soon, and don't bother looking in the yellow pages for restaurants like Darter King or Darter Cabin if you're wanting a mess of fried fish. Consider the thoughts of Aldo Leopold, who articulated for the first time the idea that all parts of an ecosystem play important roles, even if we don't recognize those roles yet. "The first rule of tinkering," he wrote, "is to keep all the parts." In Louisiana darters are a diverse group of parts.

Domestic Predators

Several species of Louisiana birds are declining because of habitat destruction and other environmental problems. A predator relatively new to the scene also threatens some. Originating in North Africa these highly efficient killers sailed with Europeans to North America and soon began wreaking havoc on wild bird populations.

Recent studies show that they kill hundreds of millions of birds in this country each year. Ironically, the culprit may be snoozing contentedly on your couch at this very moment.

Nearly 30 percent of American households have a domestic cat. The estimated number of pet cats in the United States grew from 30 million in 1970 to 60 million in 1990. The combined total of pets and free-ranging cats is likely more than 100 million. Without a doubt cats provide companionship and pleasure to many people. The downside is their instinctive predatory habits that in some cases have helped put several birds and small mammals on the endangered species list. Cats threaten the future of clapper rails, least terns, and snowy plovers in California. In the Florida Keys marsh rabbits are in trouble because of cats, and cats have driven several unique species of island mice to near extinction.

Free-ranging domestic cats also transmit diseases to wild animals. They have spread feline leukemia virus to mountain lions and may have infected endangered Florida panthers with distemper and an immune deficiency disease. Some also carry diseases that are easily transmitted to humans, including rabies and toxoplasmosis.

Feral cats in rural areas cause most of the damage. Many are dumped there by irresponsible people. However, even well-fed urban house pets will routinely kill live prey when they roam outdoors. A study in England using radio-collared cats surprised many pet owners when it revealed the deadly works of their otherwise nonchalant felines.

As cat owners, what can one do? Keep only as many pet cats as you can feed and care for. Have them neutered to eliminate inevitable and often undesirable reproduction. For the sake of your pet and wildlife, keep your cat indoors. Place bird feeders in sites that do not provide cover for cats to wait in ambush. Don't dispose of unwanted cats by releasing them in rural areas. Don't feed stray cats, and eliminate sources of food that attract them.

It is important to remember that the conservation dilemma regarding cats and wildlife is not a problem of feline behavior—after all, cats are naturally programmed with their instinctive behavior.

It is more an issue of human-induced modifications to the environment. When domesticated animals enter natural areas, conflicts will occur. Twenty-five centuries ago, a famous Greek storyteller posed a question that cuts to the heart of this matter. Aesop asked, "Who shall bell the cat?" The answer, of course, is that we all must.

Cottonmouth

While cleaning out a beaver-clogged culvert deep in a Louisiana swamp, a colleague was bitten by a cottonmouth—twice. The first bite happened as he reached into the dark pipe. Thinking it was only a briar, he reached in again and was served a second notice of trespass. As my wife would say, this sounds like a nightmare covered in molasses.

In Louisiana cottonmouths, or moccasins as they are often called, are the most frequently encountered poisonous snake. Like rattlesnakes and copperheads they are pit vipers and locate their prey with the help of infrared sensing organs in front of their eyes. Cottonmouths are semi-aquatic and spend most of their lives near water. This habit wreaks havoc with the many species of harmless watersnakes that share their habitat, as most people don't bother to discriminate between species and ruthlessly persecute them all.

Although it may not be the case in some parts of the world where aggressive, venomous snakes are abundant, the fear of snakes for many people in Louisiana goes beyond reason. It is inordinate in the extreme to their actual danger. The term "ophidiophobia" is used to describe the unreasonable fear of snakes. Research indicates that this behavior is at least partially innate, perhaps buried in a vestigial cluster of cells deep within our hypothalamus, a carryover from times and places when snakes were serious threats to humans. On the other hand, the fear also seems to be learned to some degree and socially contagious.

In the United States, less than 1 percent of all deaths by snakebites have been caused by cottonmouths. Mortality resulting from the bite of any kind of wild snake, as opposed to those kept in captivity, is extremely rare. This must, however, in no way diminish the gravitas due a poisonous snakebite should it occur. Prompt medical care is mandatory. After treatment my friend who was twice bitten still experienced considerable swelling and discomfort for a while.

Other studies suggest that unreasonable fear of danger such as snakebite can affect the health of those doing the worrying. Another friend was shot in the leg when the pistol that he carried to kill snakes accidentally discharged in the holster. These types of threats and the fact that one is exposed to substantially greater risk of being struck by lightning while standing on the bayou bank fretting about cottonmouths are the odds-on dangers of snakebite.

Dragonflies and Damselflies

Harry Potter's world includes a Department for the Regulation and Control of Magical Creatures. He does not, however, have a monopoly on beasts with mythical-sounding names. On any spring outing in Louisiana one is apt to come across a little blue dragonlet, lilypad forktail, or ebony jewelwing. Black-shouldered spinylegs, royal river cruisers, and wandering gliders are no less common. Perhaps no

Eastern Pondhawk

group of animals has more fantastic names than the order of insects called Odonata. Locally, we slight them by labeling them all dragonflies or mosquito hawks.

They are characterized by two pairs of transparent wings, elongated bodies, and multifaceted eyes. Before getting down to the species level, scientists divide the order into dragonflies and damselflies. When at rest damselflies tend to hold their wings above the body, whereas dragonflies hold their wings horizontally when perched. As aquatic larvae and adults they are ferocious predators, helping to control insect populations, such as mosquitoes and flies. Females lay eggs in or near water that hatch into nymphs with gills and hooked jaws. Often dragonflies and damselflies spend more of their lives in the aquatic stage than as adults. At some point, though, the nymph crawls up onto vegetation, sheds its old skin, pumps up new wings, and flies off to patrol the wetlands with a vengeance.

Vernacular names from around the world associate these animals with mythical beliefs. In old England they were called "devil's darning needles," and in Sweden they were known as "troll's spindles." The Norwegian name for dragonfly means "eye poker." Because they are sometimes associated with snakes, the Welsh name means "adder's servant," and in the Deep South of our country the term "snake doctor" refers to a folk belief that dragonflies can sew snakes back together if chopped in half.

At least 132 distinct species of dragonflies and damselflies are found in Louisiana wetlands. We've no need to visit the Hogwarts School of Witchcraft and Wizardry to learn about mythical creatures. Widow skimmers, common baskettails, and swamp darners are in our midst.

Mammal Dentition

If Little Red Riding Hood had prowled about the Louisiana bayous and come across a grinning possum, she would not have been nearly as dazzled by Big Bad Wolf's dentition. Possums have fifty teeth, including quite formidable canines, while the lowly wolf can only boast of forty-two. Teeth more than any other anatomical feature define many animals. A single tooth can reveal the age, sex, diet, health, and species of an animal that lived thousands of years ago.

Some fish, reptiles, amphibians, and primitive birds have teeth, but the highest level of dental complexity is found in mammals. Their teeth are composed of a core of dentin surrounded by a crown of enamel, the most durable body part known. The teeth of mammals are specialized into varieties. Incisors are found in the front of the jaw and are designed for cutting. In rodents, such as rats and beavers, and also in rabbits they grow continuously throughout the life of the animal. Frequent gnawing keeps them worn to an effective length. Canines, located behind the incisors, are spear-like and used for piercing prey and tearing flesh. Most predators have prominent canines. Premolars and molars grow in the back of the jaw and are efficient at grinding and macerating food. Grazing animals depend on them.

Not all mammals have all four types of teeth. Herbivores, animals that eat mostly vegetation, don't need canines, and thus they are absent in deer and cottontails. Deer, like cows, have no upper incisors. They nip off grass and leaves between the lower incisors and the hard upper boney palate. This fact frequently astounds novice deer hunters when they discover their trophy has no front teeth. Meat eaters, such as members of the cat family, have pronounced canines but reduced grinding teeth. Animals that eat meat and vegetation, bears and humans, for example, have teeth intermediate in development between those with more specialized diets.

Mammals usually develop two sets of teeth—milk, or deciduous, teeth and permanent teeth. Methods have been developed to determine the age of many kinds of animals by the eruption patterns and the degree of wear of different teeth. In raccoons the first permanent incisor appears at 65 days, the first molar at 78 days, and the first canine at 105 days of age. Deer are aged in yearly increments by the emergence and amount of wear on premolars and molars.

Mammals are the only class of animals with a precise number of permanent teeth in each species. Fox squirrels have 20, gray squirrels have 22, swamp rabbits have 28, cougars have 30, humans have 32 as do red bats and deer, and skunks have 34. Louisiana's natural roto-tiller, the enigmatic armadillo, has 32, and all of them are peg-like molars. Ms. Riding Hood would not be impressed.

Honeybee

For thousands of years humans have gathered honey from the hives of wild honeybees. People ate honey, concocted the alcoholic beverage "mead," and made candles from beeswax. Honey and wax were also used for medicinal purposes. None of this, however, happened in North America until Europeans arrived because honeybees are not native to the Western Hemisphere. About twenty thousand different kinds of bees are found naturally throughout the world, many types in North America, but not the honeybees that we know. Records indicate that honeybees were shipped from England to the colony of Virginia in 1622. Other shipments were made to Massachusetts around 1630. Swarms of these early colonies soon escaped and became the "wild" honeybees of America. They had spread westward to Louisiana by the Civil War as was noted by a Confederate major near Monroe in 1864. He wrote: "The troops had nothing to do but bathe in the clear water of the Ouachita and make pumpkin pies, gather wild grapes, and hunt wild honey, of which the surrounding woods furnished an abundant supply."

For centuries colonies of honeybees have been kept in wooden boxes, straw skeps, and pottery containers. In America, pioneers kept bees in "bee gums," sawed sections of hollow logs that were used as hives. The black gum tree was a favorite because of its tendency to form hollows. Collecting honey from any of these hive types usually resulted in killing the bees and loss of the colony. In the mid-1800s, a Pennsylvania minister patented a hive with movable frames that is still used today. It allowed collecting the honey without loss of the colony. Modern beekeepers in the United States produce about 70,000 tons of honey per year, while Americans consume more than 150,000 tons in the same period, resulting in an abundance of imported honey—and all because of our sweet tooth.

Wasps

Is there a grown man in Louisiana who as a boy has not, in spite of dire warnings, chunked rocks at a wasp nest and paid the dear, dear price? I doubt it, myself included.

Most stinging insects, which include wasps, bees, and ants, are close kin and placed in the order Hymenoptera. About 25,000 types of wasps are found around the world. The adults are characterized by having a narrow waist between the first and second abdominal segments. Wasps can be divided into two main groups: solitary wasps and social wasps. Solitary wasps include mud daubers, potter wasps, and digger wasps. They produce no workers and build individual nests. Mud daubers are named for their habit of building brood chambers of mud in which the female wasp stuffs paralyzed spiders as food for her larvae. Solitary wasps rarely sting humans.

Social wasps include hornets, yellow jackets, and paper wasps. They live in colonies of up to several thousand individuals that consist of males, females, and sterile workers. Social wasps build papery nests of chewed-up plant fibers. Those of yellow jackets and hornets are layered cells wrapped in a globular outer covering. Paper wasps build flat, open nests of a single comb.

Only mated queen wasps survive the winter. A dominant paper wasp queen starts the first nest of the year, often with help of other submissive queens. The first eggs develop into workers that take over nest-building and brood-rearing chores. At the end of the season a few male wasps are produced, which fertilize the over-wintering queens to renew the cycle.

Most types of social wasps aggressively defend their nest. Because the stinger is a modified egg-laying organ, only females can sting. Unlike some bees, which sacrifice their lives when they sting, wasps have a barbless stinger that can be used many times. During a sting, venom is injected into the skin of the victim and nerve endings of pain receptors are promptly stimulated. Usually, a short burst of vigorous, aerobic exercise follows almost simultaneously. If the exercise results in a murdered wasp within a 15-foot radius of its nest, the situation quickly deteriorates. Dying wasps release a pheromone that attracts revenge-minded sisters. Retreat is always a better option. About two people out of a thousand are hypersensitive to wasp stings, and an encounter can be fatal if not treated promptly, usually with epinephrine.

It's important to remember that native wasps are a spoke in the wheel of our ecosystems. A few are pollinators, and most serve to keep in check various insect populations. Aldo Leopold, the father of modern wildlife conservation, once said, "The last word in ignorance is the man who says of any plant or animal, What good is it?"

Alligator Snapping Turtle

Many years ago, we sat on the bank of Bayou de l'Outre, my grandmother and I, fishing for bream. In order to keep me more or less sedentary, she occasionally broadcast warnings of cottonmouths lurking among the cypress knees or red wasps on platter-sized nests just above my cane pole. The admonition that was most effective at keeping my toes dry, however, pertained to snapping turtles, and

Alligator Snapping Turtle

the fact that if one were so unlucky as to be subjected to the grasp of such a beast, it would not turn loose until it thundered. That and a careful examination of the cloudless sky kept my tail firmly planted on the lard bucket.

The turtle that Grandmother called a loggerhead is otherwise known as the alligator snapping turtle. It is the cousin of the smaller common snapping turtle and is easily distinguished by the three high-peaked ridges of scales on its back. The alligator snapping turtle, confined to river systems that drain into the Gulf of Mexico, is the largest North American freshwater turtle. Large males can weigh more than 200 pounds. With a carapace up to 26 inches long, a massive head, a hooked beak, and strong claws, they are formidable predators of Louisiana bayous and swamps. By scavenging and using a worm-like natural lure attached to the base of their gaping mouth, they feed on fish, crawfish, mussels, snakes, small alligators, other turtles, water birds, and mammals, and even acorns and wild grapes. They rarely bask on logs like other turtles, and only nesting females leave the water. She digs a hole and lays up to fifty eggs in the spring. At Black Bayou Lake, research has shown that raccoons destroy most of the nests. Humans and large alligators are the only predators of adult turtles. The demand for turtle meat in gourmet dishes has reduced populations throughout this turtle's historical range. The U.S. Fish and Wildlife Service considers the alligator snapping turtle a candidate for listing under the Endangered Species Act. Currently, Louisiana is the only state to permit the legal harvest of alligator snapping turtles.

When I was in college studying such things, we received word that an alligator snapping turtle had been caught near Catahoula Lake with a stone arrowhead embedded in its shell. How can that not stir the imagination? Although they have lived to be more than 70 years old in captivity, relatively little is known about their life histories as they prowl about the muddy bottoms of our bayous. Considering the number of fast food restaurants available, it would be a shame to eat the very last one.

Butterfly Puddling

It's a drinking thing, you know. We've all seen it. A group of young males get together in a vacant parking lot or down along the bayou. Hormones are involved, as are often other body fluids. The purpose of the gatherings is to drink until libidos are sufficiently enhanced to deal with any available female that might pass by. Not a pretty sight, you say? Well, actually, it is a very pretty sight. We're talking butterflies here.

Using their recoilable, straw-like proboscis, butterflies drink fluids other than nectar for moisture and minerals. A phenomenon known as puddling occurs when clusters of butterflies crowd into a small spot to drink. Only newly hatched males puddle. Older males and females do not participate. Puddling sites are rich in mineral salts and other nutrients. Mammalian urine is often the attractant. The exact purpose of puddling is not known, but scientists think that the activity may result in the formation of pheromones, a chemical used to attract mates, or other reproductive development. Not all types of butterflies puddle. Only species like swallowtails and sulphurs, whose males patrol territories, engage in puddling. Others, like hairstreaks and coppers, whose males lie in wait for females to approach, do not puddle.

You can create a butterfly puddling area in your garden. Sink a shallow bowl to ground level and fill it with moist sand. Note that the use of urine is optional, and that a sprinkling of table salt will work fine. Occasionally add a piece of fermenting fruit, such as banana or cantaloupe, as extra attractant. Then sit back and wait for the bachelor party to begin!

Peregrine Falcon

In the spring of 1996, a feathered bolt of lightning launched from the top of a skyscraper in downtown Minneapolis. During her maiden flight the young peregrine falcon tested long, pointed wings that make her species the fastest fliers on the planet. She soon learned to knock the city pigeons from the sky by sheer force of impact and returned to roost at her nesting site on top of the office building.

Peregrines are found around the world and have been worshiped by kings and sheiks for centuries as the most sought after weapon in the ancient sport of falconry. Admiring owners harness the prowess of semi-tame falcons to hunt game birds. They have also served humanity in ways other than recreation. At the beginning of World War II, the Royal Air Force trained peregrines to intercept Nazi carrier pigeons in the time-honored tradition of Caesar, King Richard I, and Otto von Bismarck. By the 1960s, peregrine populations in North America had plummeted as the pesticide DDT worked its way up through the food web into the falcons. The U.S. Fish and Wildlife Service formally listed the bird as an endangered species. DDT was eventually banned, peregrine restoration programs began, and the species slowly recovered.

The Minnesota falcon ranged farther as the summer progressed and continued to hone her hunting techniques on wild birds up to and including her size. Larger than her brother, she eventually left him and the nest site for good. Later in the autumn she began to respond to an urge with origins in her Miocene ancestors—the mysterious phenomenon we call migration. She followed the crests of north-south ridges, riding thermals to ease her journey. Traveling flocks of shorebirds and waterfowl provided ample food. By January 18, 1997, she had flown nearly 800 miles and crossed the invisible boundary between the Natural State and the Sportsman's Paradise following the sinuous Ouachita River with its attendant bayous. Barely 5 miles into Louisiana her pilgrimage ended when a person who likely claimed to be a hunter shot her.

We know these details of her life because biologists banded her in the nest on the skyscraper, and someone brought her mortally wounded body to my office. We don't know the details of the perpetrator's life. In other times when human survival depended literally on intelligence, this person may not have lived beyond adolescence. He does not think.

Ticks

They are close kin to spiders and vampires. They are uninhibited creepy-crawlies that will sneak up your britches leg to your most private of anatomical parts with the intent and sophisticated capability of doing nothing less than sucking the blood from your living torso.

Ticks do this for a living. In addition to humans, ticks parasitize many species of mammals, birds, and even some reptiles and amphibians. Being vectors of several human and animal diseases, they occasionally leave behind more than an itching bite. Lyme disease, Rocky Mountain spotted fever, tularemia, and relapsing fever are transmitted to people via ticks. Generally, tick-borne diseases are a result of a specific tick-host combination, and are geographically limited. For instance, two types of ticks are the primary transmitters of Lyme disease in the United States, and nearly 90 percent of all Lyme disease cases have been reported in northeastern states. One study there showed that most people catch it from ticks in their own back yards. Ticks are often found in leaf litter and on plants. There is a common misconception that ticks jump or fall from plants onto a host. Actually, ticks are acquired only when they are physically touched.

Most ticks have three growth stages: larva, nymph, and adult. For the organism to progress from one stage to the next, it must have a blood meal in each case. The entire lifecycle usually takes two years to complete.

If you have been randomly selected by a tick to participate in its lifecycle, some doctors recommend a specific tick removal procedure. Use fine-point tweezers to grasp the tick at the place of attachment, as close to the skin as possible. Gently pull the tick straight out. Don't crush or burn the tick, as this may release infected fluids into the bite. Don't try to smother the tick with petroleum jelly or nail polish, as the tick's oxygen will outlast your patience. Of course, you should contact your doctor if any unusual symptoms or problems associated with a tick bite develop. The wisest course of action is to always be alert to the presence of these real vampires as you complete your lifecycle.

Vultures

Writing of the carnage at Vicksburg during the Civil War, a teenage girl living 50 miles to the west near what is now West Monroe, Louisiana, made an interesting natural history observation. She stated: "we hear from the best and most direct sources that the Yankee dead lie in heaps about our entrenchments; it is horrible to relate, sickening to think, but so curious a fact that I must note it down, all the vultures have left this country, a carcass may lie for days untouched, those creatures have gone eastward in search of nobler game; how terrible is war!"

Vultures get a bad rap. At best they are thought of as nature's garbage men—not a bad label, by the way. At worst they are

Black Vulture

considered dirty, disease-carrying scavengers—not a true representation either. Vultures were once thought to be kin to birds of prey, such as hawks and eagles, but recent DNA work has revealed a much closer tie to storks, and they are now placed in that family.

Commonly called buzzards, two types of vultures are found in Louisiana. Turkey vultures are the largest, with a wingspan of 6 feet and a long, rounded tail. Black vultures have a wingspan of 4 ½ feet and a short tail. Adult turkey vultures have a red, unfeathered head. A black vulture's head is gray. With a little practice vultures can be differentiated even in flight. Black vultures appear short-winged and short-tailed, and alternately glide and flap with quick, snappy wingbeats. White patches on the wingtips are sometimes visible. Turkey vultures look long-winged and glide with a pronounced dihedral or V-shaped angle to the wings.

Both species eat carrion, and the increasing roadkills resulting from the growing number of vehicles on the nation's highways are thought to be contributing to a slight increase in vulture populations. Black vultures are more aggressive and are known to kill small animals and even weak livestock in unusual cases. Black vultures depend mostly on sight to locate their food, while turkey vultures depend on a remarkable sense of smell. Gas companies have used them to find leaks in pipelines. A strong-smelling chemical is pumped through the pipes, attracting vultures over the leak. Gas company crews just look for the soaring vultures.

Nests are usually on the ground under brush, against a log, or in vacant barns or outbuildings. In Louisiana they are commonly found in duck blinds located over water in lakes or bayous. No actual nest is constructed and two eggs are usually laid. Outside of the nesting season, vultures form communal roosts where many birds gather to spend the night in a small group of trees.

Vultures live at the apex of the ecological pyramid. As such they are more vulnerable to changes in the ecosystem than other species. Their close cousin, the California condor, is an endangered species that was almost lost to extinction because of this precarious status. If the buzzards disappear in Louisiana again, it will not likely be due to a battle but rather as a result of poor decisions that impact the environment. In this light, vultures are kin to the coal mine canaries.

Mosquitoes I

You might not think that the foothills of the Brooks Range in northern Alaska, Marsh Island on the Louisiana Gulf coast, and the Platte River in central Wyoming's high desert country have much in common, but for me there is a link. It was in these disjunct locations that I had my most memorable encounters with mosquitoes. The story line was the same whether they emerged in clouds from tundra, marsh, or enigmatic xeric wetlands. Their sheer numbers and tenacity overwhelmed any prophylactic measure involving DEET or netting and rendered all normal human activities impossible. In short, these tiny insects with only a smattering of neurons ruled big-brained humans.

The word "mosquito" is Spanish in origin and translates as "little fly." As a type of fly they go through four life stages: egg, larva, pupa, and adult. Eggs are laid in water, where they develop through the pupa stage until free-flying adults emerge. The entire lifecycle averages about a month, with the adults living only about two weeks in nature. Both male and female mosquitoes feed on plant fluids. Only the females of some species require the nutrients of a blood meal in order to produce viable eggs. Therein lies the cause of untold millions of human deaths over time. As vectors of disease-causing viruses and parasites, mosquitoes are estimated to transmit disease to more than 700 million people, resulting in 2 million deaths, annually. In Louisiana we are concerned with West Nile Virus, but most devastation is a result of yellow fever, dengue fever, and malaria in South and Central America, Africa, and parts of Asia.

For many years massive programs have been in place to control mosquitoes. Often the "cures" have proven worse than the "cause," such as the widespread application of DDT that resulted in unexpected systemic havoc. Mosquitoes and the diseases they carry have proven remarkably resilient to control efforts by mutating and developing immunities to various toxic agents. Today, the most promising techniques involve biological control, like bacteria and parasites that disrupt mosquito lifecycles. Of the more than 3,500 species found around the world, we are fortunate that only a few are capable of providing unforgettable experiences.

Mosquitoes II

Some speculate that Alexander the Great died of a stab wound. The deadly dagger, though, was not that of his Persian enemies but rather likely the proboscis of a typhoid fever- or malaria-infected mosquito. The relationship between conquest and malaria continued through the ages, destroying armies and civilians alike. Along with destruction of the Inca civilization, the Spanish brought malaria to the New World. Ironically, this invasion revealed a secret long known by the natives of Peru—the bark of a certain small tree that grew on steep Andean slopes would relieve fevers, including those of malaria. The natives called it *quinquina,* the "bark of barks." The invaders named the tree cinchona. In 1640, Jesuit priests brought this powerful malaria medicine back to Europe, but it wasn't until 1820 that two French doctors were able to isolate the potent chemical in the bark that we call quinine.

Malaria wreaked havoc on the early pioneers of Louisiana and influenced settlement patterns as many sought new homes in the piney hills, which were considered healthier places to live than land near swamps. The link between mosquitoes and malaria was not discovered until after the Civil War. Early doctors in Louisiana were slow to accept the new quinine treatment and stuck to their primitive therapies. An antebellum doctor in DeSoto Parish wrote that for malarial fevers he preferred to induce moderate bleeding, the cautious use of mercury, and blistering over the abdomen, spine, and extremities. All drugs of the period seemed to have one common ingredient. In 1832, New Orleans Charity Hospital reported expenditures for 749 gallons of wine, 24 hogsheads of claret, 1,306 gallons of whiskey, 102 gallons of gin, and 3 gallons of brandy.

By this time the demand for quinine had caused the near-extinction of the wild cinchona trees. Because of their value, Peruvian officials prohibited the exportation of the wild trees. The miracle of quinine was preserved for mankind when the Dutch government bought smuggled seeds and developed cinchona plantations on the island of Java. Since then, quinine and its synthetic clones have saved the lives of millions—including Louisiana soldiers who are fighting today in a land once conquered by Alexander the Great.

Mosquitoes III

The evolving progress in addressing public health concerns is reflected in the headlines of local newspapers. The discovery of West Nile Virus-bearing mosquitoes in a neighborhood ditch is newsworthy on every occasion, and a report of someone actually contracting the disease is front-page material. This threat-of-the-times (It only arrived on our continent about ten years ago) is real enough and of serious concern to infected victims, but it pales in comparison to the dangers our ancestors faced with a single mosquito bite. One hundred and fifty years ago while West Nile Virus was still lurking around in African jungles, pioneers in the lower Mississippi Valley were being terrorized by home-grown pathogens on a scale exponentially greater than that of today. The fear of malaria and especially yellow fever was a daily burden except during the coldest winter months. Precious children were at greatest risk. The stress was compounded by the fact that the mosquitoes' part in transmitting malaria protozoans and yellow fever virus had not yet been discovered. The fevers were thought to originate in the foul air of southern swamps, and primitive medical practices of the day usually resulted in more harm than succor to the victims. The mere use of mosquito netting in sleeping chambers could have saved many.

Headline numbers tell the story and put things in perspective. In

2007, West Nile Virus caused 124 fatalities in the United States. In just a few months of 1852, 8,000 people in New Orleans died of yellow fever. Memphis lost 6,000 citizens to yellow fever in 1878, and 20,000 died that year in the Mississippi Valley, including Louisiana. While any loss is a tragedy, today's local disease headlines are blessedly tame when compared to those of the past.

Owls

The mystique of owls has fascinated humans for centuries. Sacred to gods of Greek mythology because of their relationship with idols, owls were demonized with the rise of Christianity. They became harbingers of bad news and the associates of witches. Merlin had an owl perched on his shoulder. The dichotomy of owls as carriers of wisdom and healing or ushers of death and doom crossed cultures and continents. To Cherokee Indians owls were birds of powerful healing and wisdom, and to hear the owl call your name was a beckoning to serve your people. On the other hand, the Algonquians considered owls prophets of death and disease. For some, to hear the owl call your name meant imminent death. Even interpreters of the Bible found owls hard to understand. Different versions of the same biblical verse interchange the word "owl" with "ostrich"— quite a leap.

Found throughout the world except Antarctica and some remote islands, more than 130 kinds of owls are unique in the bird kingdom. From 5-inch elf owls of the desert Southwest to giant snowy owls of the Arctic tundra, they serve as effective predators in a variety of habitats and ecosystems. Most North American species are nocturnal and spend the hours between dusk and dawn hunting a medley of prey that may include rats, grasshoppers, mice, snakes, crawfish, and other owls. Renowned for superb night vision due to eyes with large retinas and a high concentration of light-gathering cells, owls have even more remarkable hearing. Large heads and ear openings with flat faces receive minute sounds not unlike a radar dish. These senses, along with soft, serrated wing feathers for noise-less flight, make owls unparalleled stealth predators of darkness.

Four species of owls are common in Louisiana. Great-horned owls are the largest, with a wingspan of 5 feet. Known by their size and prominent ear tufts, they nest in old hawk or crow nests or occasionally in tree cavities. Great-horned owls are among the fiercest of birds of prey and will readily attack researchers investigating their nests. Like other owls they begin nesting in late winter so their offspring will be big enough to hunt when other birds and mammals have young in the spring. The call of great-horned owls is a soft, low hooting.

Screech owls are the smallest local owls. About 10 inches long with ear tufts, they exhibit two color phases—gray and red. Screech owls readily take to nest boxes and make excellent neighbors. Their call is not a screech but a soft, mournful whinny that rises and falls down the scale.

Barred owls are the essence of our swamps. They are medium-sized owls and can be recognized by a pattern of bars running across the chest horizontally and down the belly vertically. They have brown eyes and no ear tufts. The barred owl is the typical owl of bottomland hardwood forests but occasionally visits wooded residential areas with large trees. Its call is a cacophony of weird hoots and moans that has terrified many a novice outdoorsman. The barred owl is the most common owl in Louisiana.

The least common local owl is the barn owl. It is long-legged with a white, heart-shaped face leading to the moniker "monkey-faced owl." Barn owls are birds of fields and pastures and shy away from dense forests. They are more common in agricultural areas of the state and often nest in grain silos, barns, and equipment sheds. Barn owls have a broad vocabulary of screams, hisses, clucks, and snores. The same species is found in Europe, Asia, Africa, and Australia.

For those still concerned with the forewarnings of owls, there are reported antidotes. One method is to turn your socks wrong side out. Another is to wear your hat backward . . . or you could just not give a hoot about fables.

Bayou Dan Whale

I claim to be the only person in Rocky Branch, Louisiana, with a whale in an aquarium. He shares my living room tank with two moody zebra cichlids. Other than being unfailing stimulators of conversation among visiting friends, they have little in common.

The cichlids, of course, are fish and originated in Lake Atitlan in Guatemala, and I found the whale, a mammal, in the Bayou Dan Hills in Caldwell Parish. The main difference is that the fish are alive, and the whale breached and took his last breath about 40 million years ago. Less important details include the fish being 4 inches long and the whale nearly 70 feet in length. The fish are also intact in my aquarium, but unfortunately the whale consists of only a single tail vertebra that resembles a large, brown, petrified mushroom.

The bones of *Basilosaurus cetoides,* as paleontologists dub him, have been found in a band across Louisiana, Mississippi, and Alabama where Eocene marine deposits are exposed. This layer surfaces near the community of Copenhagen in Caldwell Parish. Heavy rains occasionally erode whale bones, great white shark teeth, and coral from steep hillsides that were once the bottom of a warm, shallow sea. Whale vertebrae were so common in some areas that settlers used them as fireplace andirons and blocks to support cabins. In 1843, Dr. Richard Harlan first described this species from bones collected in Caldwell Parish at bluffs along the Ouachita River. Thinking that he had the remains of a giant ocean-going reptile, Harlan named the animal *Basilosaurus,* which means "king of the lizards." Later, another scientist found a complete skeleton in Alabama and recognized the creature as a primitive whale. An impressive specimen hangs from the rafters of the Smithsonian Institute today.

In his prime my whale had a streamlined body that resembled a sea serpent more than a modern whale. His head was 5 feet long with teeth-laden jaws that allowed him to capture and gulp down fish by the tub full. The cichlids don't seem to worry much about this though, and his surviving tail bone is a constant reminder that, indeed, "the times, they are a changin'."

Fireflies

Fireflies, also called lightning bugs, often evoke childhood memories of summer evenings sprinkled with the magic of fairy-like, flashing insects drifting about in the highest treetops. To the fireflies themselves, though, the blinking lights mean sex and supper, and not necessarily in that order.

Fireflies are woven through the mythologies of cultures as diverse as African, Chinese, and Native American. They are a type of beetle, and more than two thousand species are found in tropical and temperate regions of the world. Most kinds found in the United States are less than an inch long. Females lay eggs in the soil, which hatch in about four weeks into a larval stage. The larvae, commonly known as "glowworms," spend the summer eating other insects and snails, and over-winter in the ground. The following spring they pupate and emerge as adults.

Fireflies are intriguing because of their "fire," or bioluminescence. Sometimes called "cold light" because it produces no heat, the light is generated by the combination of several chemicals in special abdominal cells. A key ingredient is a protein called luciferin, named after Lucifer, the former "angel of light." The reaction is somewhat similar to that found in the glow sticks that kids use at Halloween, and the brightness of a single firefly is about 1/40 that of a candle.

Each species of firefly has a unique signal that they use to attract other fireflies. In one common Louisiana species the males flash about every five seconds, and the females flash about every two seconds. Potential mates are thus located by using the code. However, in the same manner, females of some species are known to locate males of other species and eat them. Surely this is cause for considerable speculation and anxiety in the "good ole boy" world of fireflies.

Thermals and Raptors

During the early part of the nineteenth century, hot air ballooning, if not entirely safe, was still all the rage. One of the adventurous occasions in bayou country was so noteworthy as to be recorded in an 1835 newspaper. The article reads: "Mr. Elliott, the aeronaut, has attempted to make an ascension in New Orleans, but the wind proved to be too strong. After seating himself in his balloon, and cutting loose, he was swept violently across the arena, knocking down several persons in his passage. The balloon next encountered a chimney top, which was overthrown by the concussion, and Mr. Elliott's thigh was broken. Part of the bricks of the chimney falling into the car, prevented the balloon from rising higher, and it was afterwards dragged over housetops and walls, and dashed against windows, till the aeronaut's hands, face and head were shockingly cut and mangled. At length, the cords of the balloon became entangled on the masts of two vessels in the river, and fortunately for Mr. Elliott, his farther flight was checked."

Mr. Elliott, not unlike his modern counterparts, had failed to perfect the use of hot air as transportation. However, one group of animals has done so in masterful style. Raptors, which include many kinds of hawks and eagles, routinely use rising currents of warm air to travel thousands of miles with minimal effort. As the sun warms the earth it creates thermals—bubbles of warm air that serve as elevators for raptors traveling north and south on their biannual migrations. Lifting off from their overnight perches in midmorning, raptors locate a thermal and spiral upward thousands of feet on near motionless wings to the top of the draft and soar toward their destination in a long, gentle glide until another thermal is found. Often many birds use the same thermal, whirling and spinning like a giant roiling pot of raptors—a phenomenon known to birdwatchers as a kettle. These aerial examples of perfect adaptation are amazing to witness and only slightly less dramatic than Mr. Elliott's solo flight in the Big Easy.

Lubber Grasshopper

I've always heard them called graveyard grasshoppers. I don't know why, but they certainly wouldn't seem out of place stick-legging around in the cemeteries of a creepy murder mystery. As the largest grasshoppers in Louisiana, the black adults are nearly 3 inches long with red and yellow racing stripes. Their wings are only half as long as their body, rendering them incapable of flight. At best they are short jumpers and slow crawlers. Technically they are called lubber grasshoppers.

We see them first in early summer as groups of nymphs that resemble miniature versions of the adults. After growing through five stages of molts, when they are known as instars, they finally morph into the adult insect. They mate during the summer, and the female digs a hole in the soil with the tip of her abdomen. There she lays her eggs in a frothy mass about 2 inches deep. The eggs over-winter in the earth and hatch in late spring. From that point on the routine workday of a lubber involves eating. Many types of

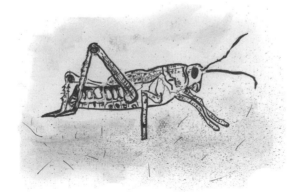

Lubber Grasshopper

leafy vegetation are devoured, especially succulent plants like jack-in-the-pulpit or those prize daylilies in your garden.

Being slow and juicy in the animal world has its disadvantages. To compensate lubbers have devised several defensive tactics to enhance their survival. Their bright coloration tends to warn others that they are not tasty snacks but rather bundles of toxic chemicals. Birds that failed to heed this warning have been fatally poisoned. They are also capable of spraying an irritant from their thorax, all the while making a loud hissing sound as air is forced through spiracles along with the foamy concoction. If this is not enough, lubbers, like most grasshoppers, can regurgitate recently chewed plant material in a dark-brown, semi-toxic liquid that my childhood mentors referred to as tobacco juice. Other keepers of mythology claim that to see a grasshopper in your dreams foretells of illness. Perhaps there lies the connection to the graveyard.

Squirrels

It has been suggested that some credit for American independence is due to squirrels, which served as abundant targets to develop a citizenry of marksmen and stew ingredients for pioneers. Perhaps they should receive more acclaim for perpetuating eastern hardwood forests as a result of their habit of burying larders of acorns, pecans, and hickory nuts. Two species of diurnal squirrels are common in Louisiana, fox squirrels and eastern gray squirrels (also referred to as cat squirrels).

Fox squirrels are generally larger, weighing up to 3 pounds. They have grayish-black backs and orangish bellies. Some local populations, such as those in the Tensas Swamp, have high numbers of melanistic or solid black individuals. Gray squirrels rarely weigh more than 2 pounds and are gray above and silver-gray below. Both species have long, bushy tails, which are used at various times for balance or as a blanket, parachute, or umbrella. Fox squirrels tend to inhabit older, open forests whereas cat squirrels prefer hardwoods with a denser mid-story canopy and ground cover. Dens are in hollow trees or leaf nests.

Squirrels usually have two breeding seasons per year. Breeding peaks in January and February and again in May and June. Three or four young are born about six weeks later and are weaned in two months. Squirrels have lived up to 15 years in captivity, but in the wild a 3-year-old squirrel is old.

The diet of squirrels consists of acorns and other mast, buds, flowers, fruit, fungi, insects, and the inner bark of trees. A single squirrel can bury several thousand acorns over a period of a few months. Some of these are recovered later by smell, but many others germinate and contribute to the reproduction of hardwood forests.

Squirrels are popular game animals, although the number of squirrel hunters in Louisiana has declined in recent years. One reason for the decline is the loss of good squirrel habitat, as pine plantations now blanket thousands of acres of once mixed pine-hardwood forests. If properly prepared, squirrel stews or sauce piquantes are considered a delicacy. If you have the opportunity to partake of such, impress your host by pointing out the species of squirrel in his pot. Fox squirrel bones are pink. Those of cat squirrels are gray. If you are so fortunate as to have the heads cooked also, point out that gray squirrels have an extra upper premolar on each side. That should validate your credentials as an American patriot and squirrel connoisseur.

Bald Eagle

What does it say about a country that shoots and poisons its national emblem into extinction? This scenario almost played out in America, at least in the lower forty-eight states. The bald eagle, that majestic raptor that adorns our currency and stands as a symbol of strength and freedom, came perilously close to disappearing throughout its range except in Alaska. The decline began a hundred years ago with human attitudes that considered all predators, including birds of prey, as vermin and targets of eradication at every opportunity. Bald eagles were randomly shot and without serious protection until the first relevant federal law was passed in 1940. Soon after World War II, a new and devastating threat emerged that swept the species to the brink of oblivion across most of the country. The widespread use of a new, highly acclaimed pesticide resulted in precipitous population declines. Birds accumulated DDT in their tissues, usually by eating contaminated fish. A few adult eagles died outright, but DDT's terminal MO caused eggshells to be fatally fragile. They broke before hatching, and with no recruitment the eagle population began to die of old age. The low point came in 1963, when only 417 nests remained in the lower 48 states. In Louisiana only a few survived in the remotest areas of coastal swamps and bayous.

Barely in time, the publication of Rachel Carson's landmark book, *Silent Spring,* in 1962 revealed the dangers of DDT and the tide turned, although it took 10 more years for Congress to ban the deadly chemical. In 1973, the Endangered Species Act was enacted, establishing the strict protective framework that led to the eagle's recovery. Now more than 10,000 nesting pairs of bald eagles are found throughout the lower 48 states with many more in Alaska. In Louisiana, bald eagles are common in the appropriate habitat, and a recent count revealed 448 nests across the state, an unbelievable number just a few years ago. This success story has resulted in the bird being removed from the list of endangered species.

It does say something about our country that we don't have to answer the first question in this story. Those citizens realizing the importance of wise conservation of our natural resources prevailed—and because of them our national emblem again soars free above us all.

Twig Girdler

A single beetle less than three-fourths of an inch long recently caused my computer to crash. My first thought was that all my valuable data were now in some black hole in another galaxy with family photos, essays, and email archives spinning around with a bunch of imprisoned light particles. Simultaneous with the computer failure, all of the smoke detectors in the house began their piercing out-of-sync chirp-whines, not unlike an imagined cat squirrel on meth. And oh yeah, the electrical transformer on the pole just outside the door exploded like a cannon shot then too. One oblivious beetle hell-bent on procreation started a chain of events that altered my day and that of several others.

The bug was a twig girdler. Her business in life is to produce more twig girdlers. She does this by chewing a V-shaped groove around the stem of a small twig, usually pecan, hickory, or oak in our area. She then lays an egg under the bark of the twig beyond the cut. The girdled twig, deprived of nutrients, quickly dies and soon falls to the ground. Clumps of brown leaves with stems that resemble a partially sharpened pencil can often be seen under yard trees after a spate of autumn breezes. The egg hatches into a larval beetle that bores deeper into the twig to feed and settles in for the winter. Pupation occurs in the cavity and a new adult beetle emerges in late summer to early fall to mate and renew the cycle. Life for the twig girdler goes on.

But insect life is hazardous, and the cycle was interrupted for at least one beetle in my yard when her egg-laden twig fell across the electrical lines, causing a dead short that blew the transformer fuse. I was more fortunate than the beetle, as summoned utility workers resolved the problem in a couple of hours. My computer data resurfaced with the fresh flow of electrons, like a revitalizing current returning to the stagnant bayou down the hill.

Frogs and Toads I

Warm spring rains often stimulate one group of resident amphibians to elicit a chorus or even a cacophony of night sounds. No fewer than twenty-seven different kinds of frogs and toads live in Louisiana. The four species discussed below are commonly heard across

Spring Peeper

the state and have calls that are distinct enough to identify the noisemaker.

Toads are distinguished from frogs by their characteristic warty tubercles and by not having webbed toes. Woodhouse's toad is the most common toad in Louisiana, found statewide except in the marshlands. Its call is a sheep-like "waaah" lasting about two seconds. This toad usually begins singing in April when air temperatures exceed 67 degrees Fahrenheit. It normally breeds in temporary pools, where the females lay long strings of eggs. The larval tadpole stage may last up to sixty days.

The next three species are members of the treefrog family. They are small with a total body length of about 1 3/8 inches, and their feet are webbed. The spring peeper's whistle-like high peep in one-second intervals may be heard on warm winter nights as well as during spring. Spring peepers are brown to olive with an X-shaped mark on the back. They are a woodland species and spend most of their time in trees. Females lay 250 to 1,000 eggs and attach them individually to vegetation in temporary pools.

Northern cricket frogs are ground-dwelling members of the treefrog family. Their skin is warty, and long, dark spots run down their gray to green backs. Their call is a sharp clicking sound, like pebbles being struck together. They may sing during any month of the year and are more prone to sing by day than any other frog. Cricket frogs are usually found near water, and studies indicate they eat insects, spiders, snails, and worms. Their eggs are laid singly or in small masses.

Chorus frogs are easily identified by their call. It is a rasping, grating chirp that sounds like a finger running over the teeth of a comb. The common name of the unique chorus frog found in Louisiana has recently been designated Cajun chorus frog. Chorus frogs are winter breeders and are usually not heard after early April. Their vocalizations are stimulated by rainfall, and they may sing when air temperatures are as cool as 44 degrees Fahrenheit. Chorus frogs are gray to brown with three dark stripes down the back and a triangular mark between the eyes. They are found in all kinds of habitats, from farm fields to forests. Females lay 500 to 1,000 eggs in soft, jellied clusters attached to vegetation in temporary pools and ditches.

Many other species of frogs and toads add their voices to the natural night sounds of the Bayou State. The presence or absence of their songs is an indicator of biodiversity and health of our ecosystems—in which we also reside.

Frogs and Toads II

A full-blooded Louisiana epicure would paddle a mile down the bayou for one. Folks around Calaveras County would trade a good used pickup for the right one. I am referring, of course, to victims of an abnormality that is on the increase—bullfrogs with extra hind legs. Imagine the limbs of just one frog, lightly battered, filling an entire skillet. Think of the chaos at Mark Twain's legendary jumping contest at the appearance of an entrant with an extra set of pistons.

Seriously, in the last several years an increasing number of frogs have been observed with severe anatomical malformations. The collective wisdom of researchers who have been working with amphibians for many years is that this is a relatively new problem. It is widespread in the upper Midwest and Northeast and seems to be spreading across the continent. Deformed frogs have been found in Louisiana. The malformations include missing feet and legs, missing eyes, one eye smaller than the other, webbing between the hind legs, club feet, and extra hind legs. Recent research has also noted internal abnormalities, such as ossification of the spine and cranium. The northern counterpart of our southern leopard frog seems to be most affected. Deformities have also been observed in the bullfrog, gray treefrog, spring peeper, wood frog, pickerel frog, and American toad. Some of these species are common in the Bayou State.

At present, suggested causes for the abnormalities are only theories. Some evidence points toward environmental contaminants. Methoprene, which is sprayed on wetlands for mosquito control and is an ingredient in cattle feed to repel flies, is suspect because it mimics a natural substance in the body (retinoic acid) responsible for early development. Some leg malformations could be caused by a natural trematode parasite that burrows into the skin of the tadpole and interferes with development. The trematode theory does not, however, answer for all the observed abnormalities. An increase in UV light due to depletion of the ozone layer is another possibility. Only additional research will affirm the true cause.

Why bother? Amphibians are indicator species. Their moist, permeable skin naturally absorbs their surroundings. They are an early warning system scattered up and down Louisiana's bayous—with warnings that we can't afford to ignore.

Black Panther

Like long fingernails drawn across a slate blackboard, some things screech across the nerves of wildlife biologists in the vein of double negatives to an English teacher. High on the list is the phenomenon of black panther reports that come in monthly. It is difficult to interpret the reason that good and honest citizens think they see animals that do not exist. Without a doubt, it is not a ruse; they fervently believe they have seen such creatures.

The large cats that once roamed the South are known as pan-

thers, cougars, pumas, mountain lions, or catamounts. They are
all the same critter, and throughout their range in North America there has never been one verifiable, documented occurrence of
a black cougar. Never. Tens of thousands of tawny, brown cougars
have been killed in this country in the last three hundred years but
nary a black one. Large black or melanistic cats do occur in the wild
in other parts of the world. Black leopards and black jaguars are
sometimes incorrectly referred to as black panthers. They live here
only in zoos.

Furthermore, many years have passed since any viable cougar
population existed in Louisiana. In recent years individual cougars have turned up, but those that weren't escaped pets were likely
wandering immigrants from growing populations in distant areas.
Radio-collared cougars have roamed several hundred miles on occasion. The odds of there ever being a sustainable population of cougars in Louisiana again are slim to none. Deer, their natural prey,
are more common than at any time in history, but tolerance
of large predators does not exist, so they will not be permitted to
live among us.

As for the black panther reports, perhaps there is a primal need
to believe in the local existence of potential predators of humans.
Maybe just the idea of a black panther prowling around the edges
of our domestic imaginations screaming on the dark moons of our
psyche is too much to give up.

Galls

In this enlightened age don't ever be misled into thinking that scientists have figured out all of the goings-on of nature, even that in our
back yards. Enigmas still abound. Consider the case of galls. Galls
are those tumor-like swellings that appear on tree stems, leaves,
bark, and occasionally flowers and roots. One of the most common types in Louisiana grows on oak trees and has the appearance
of a brown, thin-shelled golf ball. Thousands may occur on a single
tree. Insects cause most galls, and each species of insect produces

a unique gall—some are shaped like balls, others like tiny volcanoes, turbans, baskets, or bowties. Colors also vary. A tiny cynipid wasp about the size of a housefly instigates the oak galls. The process begins when the adult wasp lays an egg on a leaf or twig. The egg causes a shell of plant tissue to develop around it. When the egg hatches the larva releases hormones that direct the plant to produce tissues that form the protective gall. The gall then serves as a secure nursery for the developing larval stage of the wasp. Larvae grow in galls for several months obtaining nourishment from the inner gall tissues and emerge as adults. The adults live only about a week, during which time they mate, find a host tree, and deposit eggs to renew the cycle.

Homeowners are often concerned about unsightly, gall-laden shade trees, but they don't usually harm an otherwise healthy tree. Treatment with insecticides is not effective because the larvae are protected inside the galls. The puzzle that has flummoxed scientists is exactly how the larvae of a tiny wasp direct a giant oak to build the galls for their benefit or how each species of cynipid wasp commands the tree to produce its own uniquely shaped and colored gall . . . common examples of complex mysteries of nature yet to be solved by a species that often acts as one having all the answers.

Crawfish

In a small town near Bayou Bartholomew, there lives a man who will soon receive the very highest honor that can be obtained by a citizen of Louisiana. He is an unassuming person who lives with his wife in a modest house in a quiet suburban neighborhood. Nothing about him hints of extraordinary accomplishments or talents, except a trace of unusually abundant common sense that seeps from his conversation like dust from a farmer's britches. As it turns out, it is just this extra dose of common sense along with a big dollop of curiosity that resulted in his award.

Always the layman naturalist, he was quick to observe the excavation and construction antics of the small, gray crawfish that tunneled and mounded the loess soils in his backyard each spring. The fuss started when he innocently inquired of alleged crawfish experts a name for this critter. The men of knowledge threw at him a likely Latin moniker but asked him to supply a specimen or two for routine verification. He did this and waited. He waited for a long time. When finally pressed, the scientists muttered something about comparisons of his crawfish with known species being problematic. As requested, he provided more samples to an Arkansas professor and authority, who passed them along to an internationally known molecular biologist and crawfish expert in Utah for DNA analysis. These results yielded an amazing conclusion: the man's crawfish from a subdivision in northeast Louisiana was heretofore unknown to science.

This is a big deal in the field of biology. Outside of remote tropical rain forests or volcanoes on the bottom of the ocean, new species don't come along every day. The layman's discovery will be announced to the world in the *Proceedings of the Biological Society of Washington,* a highly prestigious journal of science. The greatest honor, though, will be in the naming of this new animal. As a tribute to the discoverer, the scientists will incorporate his surname into the new Latin name of the crawfish. In Louisiana, where crawfish is king, what greater distinction could a man desire?

Opossum

There is a wild animal in the Bayou State that is singularly unique among our native fauna. It is kin to Tasmanian devils, koala bears, and kangaroos. A persistent myth involving its reproductive habits is that it mates through the nose. This animal has opposable toes just like your thumbs and almost certainly has prowled about in your backyard. It has more teeth than any Louisiana land mammal and is even known to fake its own death when threatened. Correctly labeled the Virginia opossum, we all know it simply as the possum.

Possums are marsupials, a group of animals in which the females have a pouch where the young are suckled and raised. They are found across the eastern half of the United States and have been introduced on the West Coast. In the South they are ubiquitous in all terrestrial habitats. About the size of a short-legged house cat, the possum has long, gray hair that is used in the fur industry. A long, bare prehensile tail, a sharp-pointed nose, and naked ears combine to make this critter unmistakable. Captain John Smith of the Virginia colony wrote in 1608: "An Opassom hath a head like a Swine, and a taile like a Rat, and is the bignesse of a Cat. Under her belly shee hath a bagge, wherein she lodgeth, carrieth, and suckleth her young."

The reproductive habits of possums are unique but not as creative as the old wives' tale. Pregnancy lasts only 13 days, and when born the embryonic young weighing 1/200 of an ounce migrate to the pouch, where they remain for about 2 months. After about 100 days they leave their mother to seek their own fortunes.

Possums are omnivores and scavengers. They'll eat about anything—dead or alive. Insects, fruits, berries, birds and their eggs, carrion, even their roadkilled cousins are fair game. Except for the occasional raid on a chicken coop, possums rarely impact human activities. Indeed, they serve a critical function. Along with vultures they belong to nature's local union of sanitary engineers, that is, disposers of natural garbage. As such, they are important citizens wherever they live.

Luna Moth

This account was written by my wife, Amy, one who knows more of luna moths than is possible for mere mortal men.

In early spring nocturnal visitors arrive in bayou country. On a March evening a thumping and fluttering sound at the window announces the appearance of the first luna moth. Poems tell of moths flying into the flame of a candle, and luna moths are attracted to our lights. These pale green moths can be 4 inches across and have long, tail-like extensions of the hind-wings. They seem to be the essence of a moonlit, spring night.

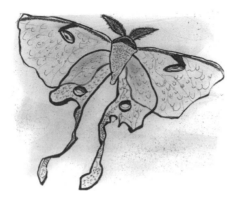

Luna Moth

The moths emerge from fragile cocoons constructed of leaves and silk. They have over-wintered in their delicate cases as pupae, and when they come out they begin their search for a mate. Since luna moths do not have mouth-parts and cannot eat, they must breed quickly. Male moths have large, feathery antennae that are sensitive to special chemicals released by the females called phero mones. Lunas have perfected their "signature" fragrance.

After mating, the female moth searches for a suitable host plant on which to lay her eggs. This is important because the next stage of the lifecycle, the caterpillar, is an eating machine. Luna moth caterpillars prefer the leaves of hardwood trees, such as birch, hickory, walnut, sweet gum, and sumac. A diverse upland hardwood forest provides this essential habitat. When the egg hatches on the host plant, the caterpillar begins to eat voraciously and grows quickly. If the caterpillar does not become a meal for a migrating warbler or a resident wren, it will build a cocoon out of leaves and silk, then pupate or rest. Luna moths in Louisiana can complete this cycle two to three times a season. The last caterpillars to pupate will become next year's earliest adults.

For many years I searched for a luna moth cocoon. I knew that a luna moth uses leaves to construct its cocoon, and that they are built on or near the ground. Since leafy things decompose quickly in the moist woods of Louisiana, I knew that finding a luna moth cocoon would be unlikely unless I found a newly emerged luna moth before it took flight. On a recent spring morning this happened. While walking a new path, I chanced upon a luna moth still pumping up its wings after emerging from its winter's sleep as a pupa. The moth had climbed to the top of a twig. I followed the twig down, and there was the empty cocoon. It was a male moth with the feathery antennae for detecting the titillating scents of the female. The cocoon, a fragile treasure made of moth silk and dogwood leaves, now hangs above my desk to remind me of the ephemeral stages of life's recurring cycles.

Ant Lion

"Doodlebug, doodlebug, your house is on fire! Come out! Come out! Wherever you are!" As a child this rhyme was my introduction to entomology, the study of insects. My mother, a south Mississippi country girl who migrated to Louisiana, instructed me to recite the passage while poking a straw into a doodlebug hole. Of course, to enhance the chances of catching this animal, one should always spit on the end of the straw first. For a 5-year-old, the educational and entertainment value of this exercise is unsurpassed.

Doodlebugs are also known as ant lions. They are members of a primitive order of insects unrelated to ants. More than two thousand species are found worldwide. Some types of ant lions have a larval stage that digs small, conical pits about the size of a quarter in sandy areas. Here the fingernail-sized, predaceous larvae armed with barbed jaws lie buried in wait for ants to pass by and slide down the slippery slopes of the pit. From an ant's perspective this critter is indeed a lion. After a period of time the larval ant lions begin the mystery of metamorphosis and change into a completely different form that resembles a dragonfly. These long-winged, feeble-flying adults are mostly nocturnal and thus rarely observed.

Pit-digging ant lions are called "doodlebugs" because of their meandering trails in the sand that resemble the "doodles" of a day-dreaming artist.

Ant lion folklore is present in cultures around the world. Rhymes and charms associated with the insect can be found in Africa, Australia, China, and the Caribbean. Mark Twain wrote in *The Adventures of Tom Sawyer:* "Doodle-bug, doodle-bug, tell me what I want to know!" One of the *Apollo 16* astronauts, while walking the surface of the moon, compared lunar features to ant lion craters and was recorded chanting, "Doodle-bug, doodle-bug, are you at home?" My guess is that he learned the verse early in life from his mother—who may have been from south Mississippi.

Blue Jay

In Louisiana don't bother looking for blue jays on Fridays. Friday is the day that all blue jays spend with the devil telling him of the bad things we did earlier in the week. Because of their occasional habit of eating the eggs of other birds, blue jays are often maligned. Even though they are striking bright-blue birds with black necklaces and white underparts, they get little respect. The famous naturalist/artist John James Audubon painted three of them near Bayou Sara in West Felilciana Parish, referred to their beauty as physical perfection, and in the next breath denigrated their general moral character by calling them rogues, thieves, knaves, pilferers, and egg suckers. The common simile "naked as a jay bird" isn't very flattering either.

The scientific name of the blue jay translates as "crested blue chattering bird." As cousins of crows they are noisy, often shrieking at intruders, such as humans, cats, snakes, or owls. Their broad vocabulary includes jeers, clicks, and gurgles. They are good mimics and may even have a sense of humor, as they frequently imitate red-shouldered hawks.

Blue jays are found east of the Rocky Mountains in Canada and the United States. In Louisiana some migrate and others are year-round residents. They are naturally birds of mixed oak and beech woods but have adapted well to urban settings. They are omnivorous, which means they eat plant and animal foods, although most of their diet is vegetable matter. Blue jays are especially fond of acorns. Like squirrels, they cache thousands of acorns in the ground, many of which later germinate to perpetuate forests. They build crude twig nests and lay three to six eggs that are often defended by the dive-bombing parents.

Other types of jays are found in North America. There are Steller's jays, scrub jays, Mexican jays, pinyon jays, gray jays, and green jays. None, though, has the character, good or bad, of our common blue jay. Even Mark Twain once said, "There's more *to* a blue jay than any other creature."

Raining Fish

Since at least the time of the grumbling Israelites, when manna descended from Heaven, peculiar things have been reported falling from the sky. Animals, especially fish, are high on the list of odd objects that have dropped out of clouds during implausible events. As an example, newspapers reported that on October 23, 1947, fish fell in the town of Marksville, Louisiana. The weather at the time was calm and it was not raining, although it was somewhat foggy. Without warning, largemouth bass, sunfish, shad, and minnows came raining down in the streets. Some of them were frozen and others merely cold, but all were said to be "fit for human consumption." A number of them struck people who happened to be on the street at the time. Fish are most commonly reported in such bizarre incidents, but other aquatic species occasionally show up, or, rather, fall down. In August 1870, a shower of salamanders hit Sacramento, California. They were apparently alive when they hit the ground. On September 7, 1953, frogs and toads fell from the sky over Leicester, Massachusetts. In May 1981, frogs fell from the sky over a city in southern Greece. The species of frog that fell was native to North

Africa. The mechanism that causes such events has long been speculated but never definitively proven. Tornadoes or water spouts are leading suspects. Tornadoes are known to have sucked up the water entirely from small ponds, and one can imagine that the smaller creatures within went along for the ride. The inadvertent hitchhikers have sometimes been carried long distances, apparently by high winds, before falling to earth, sometimes from a cloudless sky. There are many such recordings throughout history. Some reports, though, do stretch the limits of credibility. In 1877, the *New York Times* recounted that several small, live alligators fell on a farm in South Carolina. In the same year another account stated that live snakes fell over the southern part of Memphis, Tennessee. My favorite occurred in Shreveport on July 12, 1961, when carpenters working on the roof of a house had to take cover during a deluge of green peaches from an otherwise unremarkable sky. That's not far removed from manna.

Cowbird

On a summer morning at sunrise, fog wafts through the oak forest adjacent to Bayou Cocodrie as it has since the land was born of mid-continent sediments. On this morning a recent immigrant stalks her quarry in the lush understory tangle of vines and palmetto. The sleek, black invader spots her target and within seconds the work is completed, the urge to perpetuate the survival of her species satisfied for the day. Tomorrow she will hunt again for other victims.

The Latin name for the brown-headed cowbird is *Molothrus ater*, which can be translated as "black invader." A member of the blackbird family, the cowbird is an obligate brood parasite. They do not construct their own nests, but actively seek out nests of other species in which to lay their eggs. It only takes a few seconds for a female to deposit her egg in a host nest; she typically lays one egg per day and can lay thirty to forty eggs in a season. Cowbirds are known to parasitize more than two hundred species. Most birds in

Louisiana have had little contact with cowbirds until relatively recent times and haven't learned to reject cowbird eggs. The cowbird eggs usually hatch before those of the host species, and young cowbirds grow at a faster rate than the host young. As a result, cowbird chicks usurp feeding by the host parents, and in most cases the host's young starve. The host parents apparently don't recognize the cowbird chick as "foreign" and raise it in lieu of their own young.

The center of the brown-headed cowbird's range was once in the Great Plains. It is surmised that cowbirds followed roaming herds of bison, picking up insects flushed by the animals and eating seeds of prairie plants. This nomadic lifestyle may have led to its parasitic habits. Always on the move in search of food, the birds were not able to stay in one place long enough to build nests, incubate eggs, and raise their own young. Thus, brood parasitism was an adaptive strategy, allowing them to move with the herds.

Cowbirds are birds of open areas. Because of increased fragmentation of eastern forests, cowbirds have proliferated dramatically in Louisiana in the last century. One winter roost was estimated to contain more than 20 million cowbirds! Although cowbirds generally feed in open lands, they venture into the forest edge to breed and lay eggs. Roads, clearings, and other development in forested areas provide access for cowbirds and increase the vulnerability of host species. Research indicates that parasitism rates in recently logged stands of bottomland forest in Louisiana are nearly twice as high as in similar uncut stands. Parasitism rates are also related to the size of the forest block. Rates in one study averaged 20 percent of all nesting neotropical migrants in small blocks of forest as compared to 5 percent in large blocks.

At least one bird has been listed as an endangered species in part because of cowbirds. Ninety-one percent of all black-capped vireo nests at Fort Hood, Texas, were parasitized in a recent year. Consider that each female cowbird can represent the loss of thirty to forty songbird nests per year, and that their range continues to expand as a result of our ongoing landscape alterations. Such phenomena are more good examples of the connectivity of all living things, the fragility of their bonds with specific habitats, and the consequences of thoughtless meddling.

Wood Duck

No species of bird embodies the essence of Louisiana swamps and bayous more than the wood duck. Considered by many to be the most beautiful of North American waterfowl, wood ducks have been revered for centuries. Native Americans in the lower Mississippi Valley often depicted the wood duck on pottery and ceremonial pipes. Their bones have been identified at many ancient sites where prehistoric peoples lived. The first Europeans in America noticed the wood duck, and Cabeza de Vaca may have been the first to describe the species in his account of a bird he called the "royal drake" in 1527. By the seventeenth century, wood ducks were exported to Europe and bred in captivity.

Because of sport and market hunting, wood duck populations declined precipitously in the early 1900s across much of its range. All hunting of wood ducks was prohibited for several years with passage of the Migratory Bird Treaty Act in 1918. Active professional management and law enforcement led to the recovery of the species, and today populations are healthy and secure.

Wood ducks are among a small number of waterfowl that nest in tree cavities. Over much of the Louisiana landscape modern industrial forestry practices have eliminated conditions under which natural cavities develop. Today few trees reach the age necessary to form cavities, and most that do are soon cut down as undesirable. One management practice that contributes to the well-being of wood ducks is the use of artificial nest boxes. Fortunately, wood ducks readily adapt to artificial boxes placed in suitable habitat.

Thousands of nest boxes have been erected across Louisiana in recent years by conservation agencies and private citizens. Anyone with suitable habitat should consider putting up a box or two.

Several ground rules apply to ensure a successful project. First, use well-built wooden nest boxes of the proper dimensions. Boxes made of other materials have proven to be less durable, or in the case of plastic build up too much heat in Louisiana's subtropical climate. Predator guards should be used in almost all situations. Otherwise, many nests will be lost to rat snakes and raccoons. Boxes should be mounted on posts instead of trees because it's nearly impossible to construct an adequate predator guard on a tree. Finally, nest boxes should be cleaned and maintained on an annual basis. Be aware, though, that wood duck watching can become addictive.

Rattlesnakes

Few animals stir primal fear in man more than snakes. Of these reptiles, rattlesnakes seem to elicit especially powerful emotions in many people. The reason for this is not clear, and when medical facts are reviewed the fear seems unwarranted when compared to many other potential hazards that we face on a daily basis. Hospitals report snakebite of any type to be uncommon, bites from poisonous snakes even more infrequent, and human injury or death due to snakebite rare.

Three species of rattlesnake occur in Louisiana. The canebrake rattlesnake, a subspecies of the timber rattler, is one of the most impressive snakes in the state. It's heavy-bodied and obtains lengths in excess of 6 feet. Color varies from gray-brown to yellowish-gray with black crossbands and a rust-colored stripe down the back. The head has the triangular shape characteristic of all pit vipers. Canebrake rattlers are found in scattered populations across the state except marshlands and longleaf pine forests in the southeast. They are most frequently encountered in bottomland hardwood forests. Uncommon throughout their range due to habitat loss, they are not abundant even in areas of good habitat. Canebrake rattlesnakes eat rats, mice, squirrels, and rabbits. Their young are live-born instead of hatching from eggs as do many snakes, and litter sizes of seven to sixteen have been reported. Canebrake rattlesnakes have potent venom but are known to be particularly docile and often try to hide or escape if disturbed.

The pygmy rattlesnake is a small gray or tan snake with a pattern of narrow, dark bars and sometimes a pale rust-colored stripe down the back. Most are less than 2 feet long and have only a very small rattle or button. Pygmy rattlesnakes are often called "ground rattlers," but several other species of small snakes are also labeled with this misnomer. In Louisiana they are found statewide except in the southwest marshes and the delta parishes. Pygmy rattlesnakes have been reported to eat frogs, mice, lizards, and other snakes. The venom of this species is about as potent as that of the canebrake rattler, but because they can inject only a small amount, they are not as dangerous.

The eastern diamond-backed rattlesnake is rare in the state and confined to the eastern Florida parishes. It is the largest venomous snake in America, occasionally reaching 8 feet in other parts of its range. In most areas the species seeks shelter in gopher tortoise burrows. Gopher tortoises are also vanishing from Louisiana because of habitat destruction, an anecdote that makes the link between the two species more than likely.

Often maligned, rattlesnakes are a strand in the natural web of life. They help control rodent populations, and, as we are learning with many species, they probably make even more positive contributions to ecosystems that we have yet to decipher. Although they can be dangerous, the chance of injury or death from snakebite is much less than from dog bite, accidental gunshot, or even lightning strike. Consider also that if habitat destruction or indiscriminate killing exterminates rattlesnakes, we will have lost forever another facet of mystique and wildness that buffers our total domesticity on a walk through the woods.

Black Bear I

"In October, 1907, I spent a fortnight in the canebrakes of northern Louisiana . . . I was especially anxious to kill a bear in these canebrakes after the fashion of the old southern planters, who for a century past have followed the bear with horse and hound and horn in Louisiana, Mississippi and Arkansas." So wrote President Theodore Roosevelt as he described his famous adventure in the Tensas Basin that led to the legend of the "Teddy Bear."

Archaeological evidence suggests that Native Americans in Louisiana pursued bears several thousand years prior to the president's hunt. Their bones are found in association with human artifacts, traces of bear blood have been recovered from stone spear points, and their image was cast in earthen pottery. Later, during the mid-1700s, Frenchmen from Canada migrated here via the numerous waterways to ply their trade of hunting and trapping. Bears were viewed as a valuable commodity, as their hides and fat were prominent trade items. Oil rendered from bear fat was especially important in a wilderness where other forms of cooking oil were scarce. In 1806, C. C. Robin wrote, "The bears in this region are entirely frugivorous and become astonishingly fat. Individuals have been found from which 80 pots of oil have been rendered." By the mid-1800s, myriads of settlers had descended on Louisiana, and agriculture became the dominant vocation. Bears were seen as threats to crops and livestock. As a rule they were shot at every opportunity. One report indicates that in the winter of 1854–55, hunters from Prairie Mer Rouge and Prairie Jefferson killed seventy-five bears in the dense forests of Morehouse Parish. By the time President Roosevelt visited, bears were common only in the remote virgin swamps along the Tensas and Atchafalaya rivers.

Bulldozers, though, and not rifles nearly pushed the legendary Louisiana black bear into the abyss of extinction. Professional wildlife management has proven in many instances that bear populations can thrive in concert with humans in today's modern world—but only if adequate habitat exists. Extensive land clearing and draining for agriculture decimated bear habitat in the 1960s through the 1980s. More than 75 percent of the historical bottomland hardwoods in Louisiana outside of the Atchafalaya Basin has been cleared. Many of the remnant stands are in small, isolated patches that alone can't support bears. Much of the land that was cleared with the aid of government subsidies has proven to have little or no value as farmland. It is ironic but nevertheless encouraging that some government programs are now geared to reforest the same areas.

The Louisiana black bear is listed as "threatened" under the Endangered Species Act. In response to this action the Louisiana Black Bear Conservation Committee was formed, consisting of members from industry, government agencies, conservation organizations, private landowners, and others with vested interests in the issue. A goal of the committee is to perpetuate the well-being of the bear without imposing undue regulatory burdens on landowners and others who do business in bear habitat. This strategy and the success of the committee to date have been heralded across the nation as a positive example in addressing difficult conservation issues. The future of the bear in Louisiana appears bright as proactive people with commonsense solutions lead the way.

Black Bear II

In response to the decline of bears in Louisiana and the lack of information concerning their life history, the U.S. Fish and Wildlife Service in 1988 began to gather biological data that could be used to develop a management strategy to ensure the long-term survival of bear populations in the state. The study was conducted on and near the Tensas River National Wildlife Refuge. Bears were captured using baited snares, and various biological procedures, such as weighing and aging, were done. They were fitted with radio-collars for subsequent tracking and released.

Results of the study revealed that captured bears weighed from 35 to 400 pounds and ranged in age from cubs to 14.5 years. Females are generally smaller than males. Bears mate in the summer, cubs are born in winter dens in January and February, and litter sizes range from one to four. Female bears usually breed in alternate years beginning at age 3. Cubs are about 8 inches long at birth and weigh 8 to 12 ounces. They are born hairless with their eyes closed. Cubs emerge from the den with their mother in the spring and stay with her throughout the year, nursing and later eating solid food. They den with her in the winter, emerge again in the spring, and live with her until their second summer, when the family unit dissolves. The female then goes into estrus and breeds again to repeat the cycle.

The onset of denning in Louisiana occurs from late November to early January. Activity decreases considerably during this period. Most adult female bears select hollow cypress tree dens with cavity entrances 30 to 70 feet above water. Den trees are located in bayous, sloughs, and wooded flats. Adult males and juveniles of both sexes are more active during the denning period than pregnant females or females with yearlings. Most males and juveniles use ground nests in thick cover, which may include downed treetops, cane, palmetto, and vines. The ground nests are shallow, scooped-out depressions that may be bare or lined with nearby vegetation.

Radio telemetry revealed that bears are most active from dusk through dawn, although daytime activity is common. Bears often use "daybeds" under forested cover while resting. Mothers with cubs usually bed at the base of the largest tree in the area. The sow will often send the cubs up the tree if she senses danger. Bear activity revolves around the search for food, water, cover, and mates during the breeding season. Some adult males move 20 miles from their capture sites and may use up to 40,000 acres as a home range. Bears tend to follow deer trails, logging roads, and the shoreline of bayous and sloughs in their daily movements. Uncleared ditches, fence lines, and waterways are often used to cross farmland from one forested area to another.

Bears are omnivorous and depend on the natural diversity of plants and animals in bottomland hardwood forests. In Louisiana fruits of dewberries, grapes, pawpaw, persimmon, dogwood, French mulberry, devil's walking stick, and palmetto are consumed, as are a variety of insects and crops such as oats, wheat, and corn. Acorns are an especially important carbohydrate-rich food source necessary to provide fat reserves for denning.

Lessons learned from the research are more than academic. If we intend to perpetuate black bear populations in Louisiana it is vital to know how to enhance and restore their habitat. As examples, we must know that all large, hollow cypress trees should be protected as denning sites for females, that travel corridors need to be maintained and developed between isolated tracts of forest, and that oak trees should be managed for bountiful acorn production. It's not too late.

Bird Nests

Spring is a time of frenetic activity for Louisiana birds. Whether they are year-round residents or recent arrivals from wintering areas in Central and South America, most are involved in nest-building of some sort. Nests are as varied as the many species that frequent our locale and can be found from ground level to the tops of the highest trees. They can be as simple as a depression in the leaves or as complex as a finely woven bowl of spider silk and foliose lichens.

, Most people think of a typical nest as the familiar cup-shaped structure built by many songbirds. Red-tailed hawks and great blue herons, however, build platform nests of sticks and twigs; white-eyed vireos build hanging cup nests from a tree fork; barn swallows and phoebes plaster their cup nests to a vertical wall (often under bridges in our area); and Baltimore orioles weave bag-like nests suspended from branch tips.

Nest locations also vary. Killdeer lay their eggs on the bare ground of gravel parking lots or the flat roofs of buildings. Vultures build nests in hollow logs or abandoned structures. Kingfishers and bank swallows burrow into the sandy banks above our bayous to nest. Several species of woodpeckers excavate holes in dead trees. Other birds, like the tufted titmouse, bluebird, prothonotary warbler, and wood duck, also nest in cavities created by woodpeckers or other animals. Most nests are inconspicuous or camouflaged for the protection of the incubating parents and eggs or nestlings.

In constructing a nest most songbirds use a foundation of twigs interwoven with grass, strips of bark, dead leaves, pine needles, mosses, animal hairs, or feathers. Robins and wood thrushes use mud to glue their nests together. Cliff swallows build their nests entirely of mud. Chimney swifts and hummingbirds secrete sticky saliva to cement nest materials. For unknown reasons the great crested flycatcher and tufted titmouse routinely use cast-off snakeskins to line their nests. Other odd materials occasionally show up in bird nests. A five-dollar bill was found braided into a brown thrasher's nest, and a raven in Texas built a nest entirely of barbed wire. Bird nests in Louisiana vary from the 1-inch diameter ruby-throated hummingbird nest in a white oak to a 1,000-pound bald eagle nest in the fork of an ancient Atchafalaya cypress. My favorite is that of the common and perpetually busy Carolina wren, infamous for stuffing every available orifice with nest material. Not long ago researchers at Barksdale Air Force Base discovered a wren nest in a deactivated ICBM missile—a clear message from an indomitable optimist!

Feral Hog

In 1539, Hernando de Soto landed on the western Florida coast with 620 men and proceeded to march his entourage through the present-day southeastern United States in a fruitless search for gold and a passage to the Orient. Along with Hernando Cortez and Francisco Pizarro he deserves the reputation as one of the most ruthless murderers in the history of the Americas. Like the conquerors of the Aztec and the Inca, de Soto slaughtered Native Americans without remorse. Even these many victims were only a small fraction of those who succumbed to the introduced European diseases for which they had no natural resistance. Entire cultures crumbled due to depopulation.

In addition to human catastrophe, de Soto's legacy includes havoc and destruction to natural ecosystems that continues today. In order to feed his men in the wilderness, de Soto off-loaded two hundred hogs with his baggage in Florida and attempted to herd them along during his hundreds of miles of ramblings, albeit in diminishing numbers as they were consumed. Of course, many escaped to become the first feral hog populations in North America. They are proliferating in Louisiana today. Few invasive species rival feral hogs in their ability to disrupt and destroy natural areas. Hogs are omnivorous, eating any and all forms of plant and animal life. They can capture baby rabbits and young fawns. Nests of ground-nesting birds, such as quail and turkey, are eaten. Hogs eat crawfish and water-filtering mussels in shallow bayous and even large fish in drying pools. They can completely eliminate vulnerable species of native plants from an area. As voracious consumers of acorns they can greatly reduce the number of acorn-dependent animals that live in a hardwood forest, animals such as deer, squirrels, raccoons, wood ducks, and red-headed woodpeckers. For those who care about our natural areas, feral hogs should have a reputation no less infamous than that of Hernando de Soto.

Ivory-billed Woodpecker

Not far from the epicenter of the infamous New Madrid earthquake of 1811 another major shock wave occurred in 2005. This event, unlike the early nineteenth-century incident, did not cause the Mississippi River to flow backward, but the resulting upheaval in the field of natural history was just as dramatic. With unprecedented flair and cooperation the secretary of interior, secretary of agriculture, and president of The Nature Conservancy held a press conference to announce that a bird most considered extinct for more than sixty years had been rediscovered in an Arkansas swamp. Not just any bird either. This was the large, gaudy, almost mythical ivory-billed woodpecker, long sought by brigades of ornithologists—amateur and professional alike. It is hard to imagine a more stunning proclamation within the sphere of North American conservation.

In *Science Magazine*, the evidence was placed on the table: less than a dozen fleeting glimpses, some unverified sound recordings, and a very grainy video lasting only four seconds. With just this data there are reasonable skeptics. How could there not be? The Cache River swamp where the sightings took place is a large forest, but it is long and narrow. It is difficult to get around in at times, but it is by no means inaccessible. And the area is certainly not lacking in human activity. Hunters, fishermen, loggers, and even professional wildlife managers frequent the area. So why is this bird, which some of the investigators think is a single male, just now showing up? There are biological questions to answer also. The bird did not materialize from swamp vapors nor has he been hibernating for the last half century in a hollow tupelo-gum. Such a bird 15 years of age would be considered extremely old. This means that the lone survivor likely hatched of functional parents in the previous decade. Where are they? When populations of animals fall below a certain threshold they lose genetic diversity, and their ability to produce viable offspring capable of surviving and reproducing declines. Where are the other birds that harbored this critical diversity? The late James Tanner conducted the definitive research on this species in Louisiana's Tensas Swamp in the 1930s. He recorded their life history, took many photographs, and actually held a young ivory-bill in his hand. Sitting at my kitchen table, he said that when present the woodpeckers are easy to find. They are loud with raucous behavior that any amateur naturalist can detect. His observations seem to debunk the phantom image. There is also the issue of the somewhat similar and very common pileated woodpecker found throughout the South. Hundreds of reported ivory-billed woodpecker sightings turn out to be this ordinary cousin.

Of the skeptics, every one of us yearns to be proven wrong. We want to be convinced that not only does an ivory-billed woodpecker live among us, but that he has cohorts that will allow the survival of the species. We doubt with hope beyond hope that this earthquake is more than a passing tremor.

Yellowjacket Hover Fly

Mark Twain was famously quoted as saying, "Clothes make the man." Many sources stop him there and omit his following wit, which was, "Naked people have little or no influence on society." Adornments are important to animals also. A good example is that inch-long, black and yellow bedizened, hornet-looking creature that zips in to hover just in front of your face on a summer afternoon before rocketing away on a zigzag trail not unlike the cartoon roadrunner. This gaudy insect influences society also, usually in the form of terror as it buzzes threateningly inches from one's nose. Like clothes, the insect's garments are superficial, and what you see is only a ruse.

God, in a seemingly playful mood, decked out this innocent insect as an example of mimicry in the natural world. It's not a bee or a hornet, and its correct name is yellowjacket hover fly. Bees have four wings, this fly has two, not that they will ever beat slowly enough for you to count them. Bees also cannot hover like this fly. As a fly, it has no stinger and is absolutely harmless. Biologists refer to the yellow and black color theme as aposematic coloration—a bright flash of danger that tends to warn off predators. In this case, the hover fly is thought to mimic a yellowjacket. Even its loud buzz and aggressive flight is a form of mimicry.

In the South, we call this insect the "news bee" or the "good news bee" for its habit of hovering in front of a person and "giving them the news." Some say that he is actually telling you the news, while others claim that he is saying that important news will soon arrive. If one lights on your finger, which they do on occasion if offered, good luck is guaranteed. Actually, you will benefit from the presence of a yellowjacket hover fly whether it lights on your finger or not, as it goes about its important business of pollinating the flowers in your yard—a considerable influence on society.

Prothonotary Warbler

One of the most spectacular birds of our local swamps, bayous, and rivers is often called the "wild canary." The male with his golden yellow head and yellow underparts does indeed resemble a dressed-up canary. Properly known as the prothonotary warbler, this species was said to have been named by Creole settlers after the pope's legal advisor, the protonotarius, who wears yellow vestments.

The prothonotary arrives here from its wintering grounds in the tropics in March and April. It rarely strays far from water, and being a cavity nester the female usually lays her three to eight eggs in a stump hole over water. Southward migration of the surviving young and adults occurs in September and October.

Prothonotary warblers are members of a group of birds known as neotropical migrants. Neotropical migrants are birds that nest in North America and spend the nonbreeding season in Mexico, Central or South America, or the Caribbean. Many species of neotropical migrants nest in Louisiana and long-term censusing projects indicate that most are declining in numbers, some

Prothonotary Warbler

critically. Since 1966, the North American Breeding Bird Survey has shown that prothonotaries are declining at a rate of 1.5 percent annually. Reasons for the decline are complex and include loss of habitat in our area and on the wintering grounds. Predators and parasites may also be involved. If you have access to land along a local bayou, river, or lake, put a bluebird-type box up near or over the water. You may attract your very own pair of wild canaries and benefit the species as a whole.

Pollinators

What do fruit bats, wasps, lemurs, butterflies, a gecko, hummingbirds, and wasps have in common? As it turns out, along with bees they are responsible for most of the food on your table and even the centerpiece of fresh-cut flowers. They are pollinators, and without them 90 percent of flowering plants and three-quarters of staple food crops that feed the people of this planet would not exist. According to the U.S. Department of Agriculture, a pollination crisis is looming just over the horizon because of the alarming declines in wild pollinators and managed honeybees. Typical of most wholesale population declines, the causes vary. Experts agree, though, that a leading factor is the widespread use of insecticides and herbicides. Other causes include loss of habitat, parasites, invasive species, and modern agricultural practices.

Honeybees pollinate more than $10 billion worth of crops in the United States each year. However, of the hundred crops that comprise most of the world's food supply, other types of insects and wildlife pollinate at least 80 percent. Managed honeybee colonies declined from 5.9 million in the 1940s to 2.7 million in 1995.

Bats, hummingbirds, and monarch butterflies are examples of pollinators that migrate. A critical element in their lifecycles is the availability of nectar along the length of their migration route. If nectar is unavailable anywhere along the route for any reason, a part of the population may be lost. Two species of pollinating bats and thirteen species of pollinating birds are now formally considered endangered species.

To mitigate the issue, best management practices are available for pesticide applicators, land use planners, and the general public. Without awareness of the problem, they are of no value.

Gar

A sure indication of the cultural decline of Louisiana rests in the fact that not one fine restaurant offers gar balls as an entree. What has the world come to? Have all the old-time chefs in the camps along Bayou Jeansonne in Avoyelles Parish passed on?

It's not as if gar are scarce or hard to come by. Their ancestors were already abundant and cruising the cycad-draped bayous in the early Mesozoic period 240 million years ago. These air-gulping fish evolved into the four species of gar that are found throughout the state. The spotted, shortnose, longnose, and alligator gar are still considered primitive fishes. Their heads consist of bony plates and their long, tooth-studded jaws support highly predaceous feeding habits. Early French explorers called them *poison d'armee*, or armored fish, for the tough, triangular ganoid scales that protect their long, slender bodies. The scales are so hard that they were used by Native Americans for arrowheads and by farming pioneers to sheathe wooden plows.

Because of a unique physiological adaptation gar can survive in stagnant, oxygen-devoid waters. A highly vascularized swim bladder is used as a supplementary respiratory organ and allows the fish to breathe by swallowing air. This explains their peculiar habit of frequently breaching the surface of the water, especially on hot summer days when oxygen levels are low.

Gar spawn in the spring in large aggregations. Eggs are deposited at random in shallow water and receive no care from the adults. Alligator gar can grow to be 10 feet long and weigh more than 300 pounds. They are by far the largest freshwater fish in Louisiana waters.

As is the case with most predators, gar are usually considered nuisances because they compete with man for other species. For many years fisheries biologists sought ways to exterminate gar, including elaborate shocking devices, traps, and nets. They have no doubt been reduced in numbers, especially the larger individuals. What is not clear is the role that gar play in a natural aquatic ecosystem as they swim at the top of the food web. It is likely significant. In addition to controlling populations of other fish, they are known to be intermediate hosts in the parasitic stages of some freshwater mussels. Mussels are the natural filtration system in our bayous and rivers.

And now for the recipe. Hull out one very fresh gar. A hatchet works well. Debone and grind the meat in a food processor. Mix with bread crumbs, lots of chopped onions, and spices to taste. Make into balls or patties and deep-fry in hot grease until golden brown. Must be served while hot.

P.S. Never eat gar eggs. At one time they were hung in barns as rat poison in Avoyelles Parish.

Mussels

Freshwater mussels are a little known but critical component of the biodiversity of Louisiana bayous, streams, and rivers. Related to the much sought after oysters of the coastal area, freshwater mussels in Louisiana are not usually consumed by people today. Such was not always the case, however, as Native Americans routinely harvested large amounts of this high-protein food. Piles of discarded shells, or middens, still mark the campsites of prehistoric peoples across the state.

Early in the twentieth century, hundreds of factories were scattered up and down the Mississippi River utilizing freshwater mussel shells to make pearl buttons. Thousands of tons of shells were collected by commercial harvesters and sold to the button factories, which cut, drilled, and polished the pearlescent shell into buttons. Development of the plastics industry sounded the death knell for the button factories, as inexpensive plastic buttons soon replaced those made from mother-of-pearl. A renewed interest in freshwater mussels occurred when Japanese cultured pearl research revealed that tiny pieces of American mussel shells seeded into oysters made the ideal nuclei around which mother-of-pearl developed.

Ecologically, mussels are critical to many aquatic ecosystems as filters of suspended particles. They feed by siphoning water through hair-like structures called cilia and sorting out plankton and organic materials. Some researchers have found that healthy mussel populations can actually filter the entire volume of a stream in a short period of time. In doing so, they are vulnerable to pollutants from agricultural and industrial runoff. Chemicals, pesticides, and erosional sediments from poor logging and farming practices kill many mussel beds. Water quality soon deteriorates even further as the natural filters are eliminated.

About sixty-five species of mussels inhabit Louisiana waters. Three are considered endangered. Because of their sensitivity to environmental degradation, mussels are considered indicator species. When the health of a stream's mussel population declines, it usually means that other native plants and animals in the same ecosystem will soon be disrupted—and so it goes right on up the line until humans are the ones impacted.

Chimney Swift

If there were an occasion to choose the most martial of all bird species, I would not pick the majestic eagle with its fierce countenance or the falcon renowned for its hunting prowess. Instead, my vote would go to the 5-inch chimney swift. Do not be misled by their high-pitched, twittering calls sounding like the staccato noises emanating from a roomful of adolescent girls. These birds can fly! Though sooty-gray in color, they are the Blue Angels of the bird world. Wingtip to wingtip in perfect formation, they zip through the Louisiana summer sky twisting and turning with stiff, rapid wingbeats and graceful glides. They never perch on wires to rest like swallows because their tiny feet can only cling to vertical surfaces. Only hummingbirds, close relatives who function as avian helicopters, can rival the extent of their aerial maneuvers.

Chimney swifts nest throughout the eastern half of the United States and migrate in autumn to spend the winters in the northwestern quarter of South America. Originally, they nested in large, hollow trees, but because such habitat is now uncommon, most nesting today occurs in open chimneys. The twig nests are glued to the vertical walls with their sticky saliva. Usually there will be only one active nest per chimney. Populations of chimney swifts are in general decline because of loss of natural habitat and fewer open chimneys.

Chimney swifts are born aces. Two parent birds and their off-spring can down thousands of flying insects every day. Their targets are mosquitoes, gnats, and termites. So if you hear a noise in your chimney that sounds like far-away thunder or perhaps muffled cannon fire, consider yourself fortunate. It is only the wingbeats of your private air force on patrol.

Alligator

I have come to the conclusion that alligators don't travel well. It's an opinion based on several incidents that have occurred over the years in my dealings with these survivors from the age of dinosaurs. Historically alligators were found throughout Louisiana but were always more abundant in the coastal marshes and along the major river systems. Beginning in the late nineteenth century, market hunting for their valuable skins decimated wild populations to the point that biologists feared the species might become extinct. For that reason all harvest of alligators was banned in 1963. Within ten years alligator populations recovered dramatically, and the added shelter of the Endangered Species Act in 1973 enhanced their protection. The success story of alligators since that time reflects the wisdom of sound wildlife management practices. Today populations across the state are healthy, and thousands are harvested annually during highly controlled, sustainable trapping seasons.

As an employee of the U.S. Fish and Wildlife Service I was often involved in alligator-related activities, including research that entailed the capture and tagging of thousands of alligators in the coastal marsh, transporting them to reestablish populations, and responding to nuisance alligator complaints. These experiences have led me to believe that alligators don't travel well. Case in point: not long ago a 7-footer was captured at the state fish hatchery near Monroe. He had been enjoying the easy pickings of grain-fed fish in the ponds, much to the dismay of the hatchery manager. I offered to relocate the opportunist to Handy Brake National Wildlife Refuge and sent two men to do the job. When they arrived at the hatchery, the culprit was tied up and his mouth had been taped shut. As they loaded him in the back of the pickup truck, the hatchery manager casually mentioned that he was indeed a fast booger and in fact had initiated a few tense moments the preceding day, which led to a quick trip to the emergency room for one of their biologists.

Alligator

Thankfully, it was not too serious. Taking this occupational hazard in stride, our men headed north on Highway 165. They were somewhere south of the Sterlington intersection when the driver noticed in his rearview mirror the alligator, untied and untaped, slither over the tailgate at 55 mph into four lanes of traffic. He later made the observation that people rarely stop to assist a motorist in trouble—but for an alligator, they stop. There was a minor traffic jam and considerable commotion before the escapee, remarkably intact, was apprehended and sent on his way.

This story is very similar to one that happened several years ago as I was transporting a load of alligators from south Louisiana to northwest Arkansas. In the piney hills just above Hamburg, Arkansas, the same MO was used for the escape, and I found myself sitting astraddle the offender on the yellow stripe in the middle of the road. I could barely hold him in place. I certainly could not get him back to the vehicle alone, and the passing pulpwood truck driver would have none of it. During this same large relocation effort involving hundreds of alligators and several states, a friend watched an 11-footer escape his bonds Houdini-like, rise up over the back of his truck cab, devour his radio antenna, and depart the vehicle heading back south down I-49 near Jackson, Mississippi. I have also discovered that they don't even like to ride in boats. On a dark night in the Lacassine Pool a large and temporarily captured individual dismantled our artificial lights, crunched a fine 35 mm camera, and treed us on top of the airboat engine cowling. Could they be trying to tell us something?

Shorebirds

During the hottest dog days of early August few of us in the Bayou State expect a wave of visitors from the Arctic tundra. For a group of birds known collectively as shorebirds, however, this seemingly irrational behavior is quite the norm. Tens of thousands in small, tight flocks come wheeling and whirling out of the northern sky seeking our remnant wetlands.

Several species of shorebirds are found in Louisiana. Most are small wading birds associated with mudflats and shorelines, where they feed almost exclusively on insect-like invertebrates. They include several types of sandpipers, plovers, and their kin. Only two species are known to nest here, the most common being the familiar killdeer. Many shorebirds nest in Arctic regions and conduct long-distance migrations to winter in Central and South America. Some sanderlings fly 22,000 miles annually between the Arctic Circle and Patagonia.

During the late 1800s and early 1900s, shorebirds were extensively hunted for the market and suffered major population declines similar to those of waterfowl. Since 1916, with the passage of the Migratory Bird Treaty Act, hunting has been illegal for all but two shorebirds, Wilson's snipe and the American woodcock. Hunting of these two species is now carefully regulated.

Today the greatest threat to shorebirds is the loss and degradation of habitats on migration and wintering areas. Wetlands in the United States have decreased from an estimated 200 million acres to about 100 million acres. Similar losses have occurred in areas of Central and South America. Shorebirds have very narrow habitat requirements during migration, when they must find adequate feeding areas in order to fuel up for long flights. As many shorebirds observed in Louisiana are in the midst of migration, management techniques here consist of discing and flooding shallow water areas on refuges to provide critical habitat during this usually droughty time of year. We may question the rationale of leaving a cool climate to visit Louisiana in mid-August, but keep in mind that most are just passing through.

Armadillo

For several months I conducted a highly unscientific survey to determine which of our native wild mammals deserves the dubious title of Roadkill King of Louisiana. Though inconclusive, my results indicate that the ubiquitous armadillo ranks near the top, perhaps second only to possums, whose primitive neurons have yet to grasp the physics of speeding Pontiacs.

A wise wildlife professor once taught that if a particular species of animal is commonly killed on the highways, it usually indicates a high and sustainable population of that animal in spite of the roadkill mortality. Such is the case with armadillos throughout our state. However, this was not the situation until recently. Armadillos migrated from south Texas into southwest Louisiana in the early 1900s and didn't become common until the 1950s. Properly known as the nine-banded armadillo for the armor-like flexible bands protecting its back and sides, this warm-blooded mammal is one of the most peculiar of our fauna.

Often maligned for burrowing in gardens and flowerbeds, armadillos consume large quantities of harmful insects, as documented by research into their food habits. In one study in Kisatchie National Forest, beetles and their larvae comprised nearly half their diet. More recent work proves that armadillos eat vast numbers of fire ants, and their burrows provide habitat for different types of wildlife.

Breeding usually occurs in the summer and delayed implantation postpones birth until spring. Four identical quadruplets are al-

Armadillo

ways born, resulting from a single fertilized egg that divides twice. This unusual reproductive feature ensures that all four young in any litter are the same sex and genetic carbon copies.

On a more ominous note, armadillos are the only animals other than man known to harbor the bacteria that cause Hansen's disease, once known as leprosy. The likelihood of transmission of Hansen's disease from armadillos to humans is, however, very low. Researchers determined that while the leprosy bacterium is common in some Louisiana armadillo populations, there is no cause for alarm for the average citizen. In fact, the greatest harm likely to come your way from an armadillo is the cost of a front-end alignment should you meet one on the highway.

Pied-billed Grebe

The degree-toting ornithologists call her pied-billed grebe, but on the bayou she answers to "di-dipper," or "hell-diver" on a bad day. For her the water's surface is only an interim point in space and time. That she spends precious few moments there is an aggravation for birdwatchers and boys with BB guns, not to mention the serious predators, be they finned or feathered. Departing the planar ecotone dividing air and water is more graceful if she chooses the denser of the two mediums. The choreography of her downward dive is faster than the eye can appreciate. A slow-motion camera would reveal a slight rise of her bow, a cock of her thin, bowed neck like the readying of a hammer on a dueling pistol, and a forward-tipping plunge, the curtain falling with a vanishing flash of white tail feathers and a circle of ripples. While this disappearing act is nothing short of magic, her transition from bayou surface to flight is nothing short of embarrassing. Her body barely meets the minimum specifications to function as a viable airfoil, mainly because of stubby wings that seem to be God's afterthought. Lack of horsepower is also an issue. The decision to fly is made only with the greatest reluctance, and even then she would be better off to di-dip or hell-dive. With a burst of energy not commensurate with altitude gained she begins

Pied-billed Grebe

a long, pitter-pattering run across the pond that often as not ends in a dive because of a misjudgment in runway length. If the takeoff is not aborted, gravity is eventually overcome and flight occurs, albeit tentatively. Such is life for the pied-billed grebe, a bird more at home in the realm of bluegills than the ether of bluebirds.

Snake Myths

The myths that surround Louisiana snakes are almost as amazing as the irrational fear that many people have of these reptiles. It is likely that more people injure themselves trying to kill harmless snakes than are harmed by poisonous ones. The myths are just as foolish.

One persistent myth involves the legendary hoop snake. Its proponents claim that this snake takes its tail into its mouth, forms a hoop, and either rolls away from danger or attacks unsuspecting people. Documentation of this behavior is anecdotal at best. Another tale pertains to the stinging snake, which has a stinger at the end of its tail that is readily applied to slow humans. Often the stinging snake and the hoop snake are combined into one formidable comic book creature. In truth, the perfectly harmless and hoopless mud snake is the species usually credited with these characteristics.

The coachwhip, a large, uncommon snake of the hill parishes, purportedly whips people to death with its long, whip-like tail. The scales on its tail do remotely resemble the braided leather of a whip, but bodies resulting from such attacks are scarce.

At least one myth is downright dangerous. Those who believe that cottonmouths cannot bite underwater best not miss a life insurance premium if they decide to test the issue. Another is that coral snakes have to chew to inject their venom. Snakes don't and can't chew. Coral snakes bite and hold on to inject their venom. Also, snakes will readily cross a hemp or horsehair rope in spite of what some folks think. And snakes aren't slimy. Their skin is normally quite dry and mostly smooth.

A recent study revealed that some fear of snakes might be hardwired into humans through evolutionary processes that favored mammals who avoided reptiles. So perhaps people do have an excuse for being terrified of creatures 1/150 their weight with pea-size brains—or is this a myth too?

Geese I

Writers have often tried to express the feelings stirred in humans by skeins of wild geese as their haunting cries drift down from blustery skies. Until about thirty-five years ago people in northeast Louisiana usually experienced these feelings only in October and November and again in the spring as geese passed over during their seasonal migrations. Occasionally geese would make a brief stopover in a soggy soybean field or along a bayou sandbar before continuing on.

Five species of geese passed through in the autumn on their way to lush wintering grounds in coastal marshes. Snow geese in blue and white color phases were and continue to be the most numerous. The blue phase was once thought to be a separate species, but research proved it to be a color morph. Ross's goose, a miniature version of the white-phase snow goose, is barely larger than a mallard duck. Canada geese, once abundant, declined so drastically in Louisiana by the 1960s that hunting of the species was curtailed. The establishment of many new waterfowl refuges in the upper Mississippi Valley "short-stopped" many Canada geese on their way to traditional wintering areas in Louisiana. The unnatural situation resulted in unexpected problems, including disease and crop damage, and was eventually discouraged. Gradually, Canada geese are reestablishing historic migration routes and wintering areas in south Louisiana. The cackling goose, recently adorned with full species status, is a diminutive copycat of the Canada goose. White-fronted geese, known locally as "specklebellies," are second in abundance to snow geese. Their distinctive high-pitched cackle identifies the species as it flies overhead.

Unlike most ducks, geese usually nest in the desolate windswept wilderness of the high Arctic. The breeding grounds of snow geese are so remote that they were unknown to science until 1929. To some degree, the remoteness has protected geese from human activities that often conflict with the well-being of wildlife. Many ducks, for instance, nest on Midwestern prairie sites that are intensively farmed.

Beginning in the mid-1960s, two dramatic changes occurred in northeast Louisiana. One spawned the other. The first was the advent of large-scale rice farming. The second was that geese no longer just passed through; they began to over-winter in the region. Soon nearly 70,000 acres of rice were planted annually in Morehouse and Richland parishes, and more than a quarter million geese wintered where only a few years earlier none stayed. The magnitude of this change in wildlife population dynamics is unparalleled in Louisiana's recent history. The long-term consequences are complex and unknown. Have these geese been inadvertently short-stopped a bit farther down their ancestral highway in the sky? Can this adaptation to man's tinkering with the environment be considered natural? Have the coastal marshes degraded to the point where they and the south Louisiana rice lands cannot support present goose populations even if they all did migrate to the coast? The answers are as ephemeral as the north winds of an arctic cold front, but one thing is certain—the plaintive calls of wild geese have become common in northeast Louisiana . . . at least in the short term.

Geese II

The source of mysteries is not limited to the likes of Tony Hillerman or P. D. James. Nature also serves up some perplexing whodunits from time to time. On the evening of January 25, 1983, several people called the Lacassine National Wildlife Refuge in southwest Louisiana where I worked to report unusual snow goose mortality near Jennings. As the regional wildlife disease biologist I was responsible for looking into the matter. The affected site was a rural farming area about 2 miles wide and 5 miles long. Local residents reported first seeing dead geese on the morning of the 25th. Birds were found in yards, roads, and ditches as well as in open fields. I saw about fifty dead geese scattered at random lying belly up. They made indentations in the wet fields where they struck the ground, indicating that they became incapacitated while flying and fell from the sky. I estimated the total mortality to be two hundred to three hundred.

I collected and necropsied several of the geese. No lesions characteristic of infectious diseases were found although all exhibited free blood in the heart and lung cavities. A definitive cause of death could not be established, but the pattern of mortality seemed to rule out diseases, parasites, or poisoning. No known waterfowl disease or parasite causes such a rapid death, and mortality from something like pesticide poisoning might be expected to emanate from a central focus where contact with the agent occurred. Such was not the case.

The National Weather Service station at Lake Charles and local residents reported heavy thunderstorms in the area the night before the first dead geese were seen. Although the carcasses showed no signs of lightning or hail strikes, I concluded that the mortality was weather related. Severe thunderstorms are known to spawn powerful updrafts capable of lifting aircraft thousands of feet. Coastal thunderheads often tower to 7 miles above the Earth. The barometric pressure and oxygen content of air at these heights are greatly reduced. It is feasible that a flock of geese caught in an updraft and carried rapidly to such heights will experience respiratory and circulatory problems similar to those found in the necropsied birds. This theory would also explain the distribution of carcasses, as geese succumbed and fell out of the storm at slightly different times and places.

Geese have been observed flying at extreme altitudes in places such as the Himalayas. These heights were probably achieved over a period of hours, thus allowing time for physiological adjustment. The Louisiana geese were likely carried aloft in a matter of only a few minutes, which precluded their acclimation and resulted in mortality. So, nature mystery solved . . . at least in theory.

Deer I

Every year in late spring Louisiana wildlife officials begin receiving reports of abandoned deer fawns from concerned citizens. Often young fawns are observed alone with their mothers nowhere in sight, which leads to the almost always erroneous conclusion that the fawn has been abandoned. Exceptions occur, such as when the doe is known to have been struck by a vehicle, but they are uncommon. Problems occur when well-intentioned people attempt to rescue the apparent orphans. Several studies, some using radio telemetry to track deer movements, have shown that it is very common and natural for does to leave their fawns for extended periods during the day to feed. In these situations the does always return and have no problem locating their young. Even if the fawns move while they are gone, their mothers have no trouble tracking them down using their keen sense of smell. Capturing fawns is almost always counterproductive from a natural standpoint. They are rarely orphaned, and it is very difficult to successfully reintroduce pen-raised fawns to the wild. It is also against the law.

Deer II

Even the dog days of summer can't dissuade many Louisiana hunters from dreaming of frosty autumn mornings and the chance to bag a trophy buck. Much of the appeal involves the boney appendages that grow from the skull of male white-tailed deer. Bigger is better. If you want to make a biologist cringe, refer to these prized objects of desire as "horns." They are not, but rather are correctly termed "antlers." True horns consist of a core of dermal bone covered by a horny epidermal sheath. The sheath is the actual horn, and they are not usually shed. Cows have horns. Antlers are branched structures of bone characteristic of the deer family and are shed annually.

Antler growth of deer in Louisiana usually begins in April or May. The antlers are covered by skin and hair, sometimes called velvet, until they mature in the early autumn. Mature antlers are about 60 percent mineral and 40 percent organic matter. Velvet is shed and antlers are polished in most bucks by mid-October. Antlers are shed or cast as early as January, but most are retained through mid-March.

The entire antler growth-development-casting cycle is directly tied to seasonal fluctuations in day-length, or photoperiodicity. Growth begins as day-lengths increase, and physiological changes that lead to antler shedding occur as day-lengths decrease. When deer are subjected to artificial light sources, their cycles can be shifted out of phase. As would be expected, deer in the tropics where day-length is constant exhibit antler development cycles at various times of the year. These cycles do not occur in unison.

It has long been recognized that antler development in white-tailed deer is a function of at least two independent factors: age and level of nutrition. Genetics is also important. Overall, older bucks have larger antlers up to a point. Research shows that dietary energy, protein, and minerals are critical. In most situations these variables are tied directly to the soil type in a given area. Fertile soils produce deer foods high in energy and protein. Infertile soils don't. This means that the greatest potential for large-antlered deer in the Bayou State lies in the rich alluvial soils of the Mississippi and Red River floodplains. However, trophies can and do occur in other areas in response to special conditions. An entire industry has developed around the nutritional aspect of antler development. Dietary supplements in the form of mineral blocks, high-protein feed, and purported miracle clovers sell like hot cakes to hunters in search of the perfect wall-hanger. Many are of dubious value.

The science of deer antler growth and development is quite advanced. The knowledge vacuum lies in the arena of their lure and intrigue to humans.

3
MODI OPERANDI
Methods of Functioning or Operating

Observed a gator killed spike buck in Streeter's Canals. He was
about half eaten and had numerous tooth puncture marks over carcass.
A 10' gator was sunning about 200' down canal.
   —KO Field Diary, 27 October 1982—Lacassine NWR

Hurricane Katrina hits New Orleans and lower bayou country.
   —KO Field Diary, 29 August 2005

The natural world of bayou country is driven by a myriad of processes, all a blend of biotic and abiotic components with their deep-
est roots in the energy from our closest star. Photosynthesis, respi-
ration, and meiosis are no more important to life than the averages
and extremes of weather that steer plant succession. Migration, hibernation, predator/prey relationships, and disease are bound in the
physics of time. Survival of individual plants and animals is connected to survival of populations is connected to habitat quality is
connected to soil fertility. The living and nonliving parts are insep-
arable, a concept often ignored by citizens of a consumer-oriented
society.

Connections

If I could choose one concept to leave with readers of this book, it would be that of "connections"—links between humans and the natural world. Connections can also be links between plants or animals and their habitats. Everyone knows that fish need water, but few realize that many prairie plants need fire to survive. Spanish moss needs clean air, spring peepers need seasonal pools, and mud daubers need mud. Connections can be broad—all animals depend on plants either directly or indirectly, or very specific—red-cockaded woodpeckers need old pine trees infected with the fungus that causes red heart disease in the tree. Rare indigo snakes need the burrows of endangered gopher tortoises in southeast Louisiana. In the hill country of north Louisiana beech drops grow only under the disappearing beech trees.

As a species and as individuals there is little that we do that does not impact other living things by affecting connections. The consequences of our actions on other life forms can be negative or beneficial. The scale can vary from global extinction of a species to preservation of entire ecosystems. Clearing bottomland hardwood forests in the lower Mississippi Valley ensured the demise of ivory-billed woodpeckers, but conservation efforts have likely preserved the greater Yellowstone ecosystem. The same action can be fatal to individuals yet essential to the well-being of a whole population, as in the use of hunting as a tool to maintain healthy deer herds. Most detrimental impacts are the result of cumulative actions. The fact that I drive a gasoline-powered truck each day doesn't impact caribou in northern Alaska. The fact that 50 million people drive similar vehicles may seriously affect caribou as the demand for oil and its associated activities in the fragile Arctic displace the herds.

Knowing the connections is important. It is even more important to realize that we have not yet figured out most of the connections, and that many will always elude our comprehension. These in particular are no less significant to the players involved. Recognizing that unknown connections exist is as important as proven facts in making daily decisions that impact the natural world.

Tree Fall

Deep in the D'Arbonne Swamp just on the bayou side of Wolf Brake a giant, forked willow oak split at the confluence of the two trunks and crashed to the forest floor. Barring thunder and gunshot it was probably the loudest sound in that neck of the woods in many a year. The odds are good that no humans were around to hear it, but certainly nearby wildlife went to red alert at the first crack. A scenario in which a doe in an adjacent thicket snorted and headed for the hills, a fox squirrel bailed out of a leaf nest, and a barred owl flushed indignantly from a cavity in the doomed tree is not unrealistic.

When I found it, the tree had recently fallen, probably as a result of the fringe winds of a hurricane that destroyed entire forests 150 miles to the south. Here the impact was less dramatic but just as crucial to the survival of a forest. Where the tree once stood something absent for a hundred years suddenly appeared on the ground. This magic elixir spilled onto fallen leaves, splashed into once dark crevices, and kick-started a complex chemical reaction that would ensure that a bottomland hardwood forest would remain in D'Arbonne Swamp. This fertilizer of fertilization, this gatekeeper of reproduction, was sunlight. For a century past the oak had shaded the ground underneath, enforcing a death sentence on the thousands of acorns that had fallen from its branches in the many autumns. The seedlings of oaks, like many swamp species, are shade intolerant and will never grow into mature trees while imprisoned in the depths of shade. Such forests reproduce naturally only in the sunlight-splattered gaps created when large trees fall to the earth. Then, long suppressed seedlings grow vigorously toward the source of light around which our planet orbits. The sound of the falling tree was the birthing cry of a forest.

Lightning

A recent night of thunderstorms temporarily assuaged the tree-killing drought. By mid-morning of the following day, plants were displaying the reprieve in leaves with restored turgor pressure after weeks in a state of progressive wilt. Two inches of rain brought by the storm without a doubt saved the lives of some critically stressed trees. However, as is often the case with natural events, there were winners and losers in the passage of the midnight storm. Within a hundred feet of my house in the woods three bolts of lightning came to earth in rapid succession. One killed the computer that harbored the Bayou-Diversity programs. Backups aside, it was only a mundane tragedy. For me the greatest losses were two white oaks and a mockernut hickory that germinated when Ulysses S. Grant was president and whose sap of life was boiled away at the speed of light. The explosions that rattled our bedstead resulted from after-the-fact acoustic waves when the air along the lightning's path was heated to 36,000°F—three times the temperature of the sun's surface. The eyelid-penetrating flash of light occurred on the return stroke, the part of lightning discharge that is visible. In about 30 millionths of a second the trees' sap, being a poor conductor, was boiled to steam as bark exploded along the long vertical seam. The prognosis is not good. Even if they manage to survive the winter, the trees' wounds are ripe targets for bacteria and fungi that result in life-threatening decay. In medieval Europe "aeromancy" was a term used to describe the prediction of future events based on the observation of weather conditions. My divination of the recent storm forecasts a trip to the computer store and a big pile of firewood.

Backwater

Few people in Maine, Wyoming, or California can relate to the term "backwater" like those who live in bayou country. It refers to the natural, cyclic overflow of rivers and bayous that inundates areas characterized by bottomland hardwood vegetation. Backwater generally occurs in winter or spring in response to heavy, seasonal precipitation on local watersheds or as far away as the upper tributaries of the Mississippi River. The key word in this definition is "natural" because backwater has created much of the land that we know and continues to shape the flora and fauna.

Backwater dictates the type of plants that grow in overflow areas by replenishing shallow water tables to ensure that only species adapted to live in wetlands can survive. Pine seedlings frequently invade swamps during dry cycles only to be killed when the floods return. The rising and falling waters disperse floating fruits and seeds of mayhaw, overcup oak, water hickory, and cypress to provide diversity throughout the ecosystem.

From longnose gar to largemouth bass, backwater is the key to many fisheries by providing critical spawning habitat. Backwater allows the temporary passage of fish from one oxbow lake to another, again ensuring diversity down to the genetic level. Native terrestrial wildlife have adapted to the floods, routinely following the water in and out of the swamps. Slowly rising waters cause few problems for most species if suitable habitat is available in nearby uplands. Deer along the Mississippi River give birth to fawns up to two months later than those in nearby hills, perhaps to avoid backwater at a critical time.

The most important function of backwater is likely the infusion of nutrients to fuel the system from the bottom up. Several hundred thousand acres of former backwater areas in Louisiana never or rarely flood because of levees, ditches, pumps, and dams. Most have been converted to agriculture. Even in remaining forested areas the cycle is broken, and the land is never as productive. Nutrient-deficient plants eventually produce less fruit, acorns, and browse, lowering the carrying capacity of the deer herd. Lack of flooding results in fewer fish and crawfish to support great blue herons, raccoons, and otters. The absence of backwater means less seed and animal dispersal and thus less diversity. When diversity decreases to a finite point, ecosystems implode and cease to exist as a sustainable unit.

For thousands of years humans adapted to backwater and even exploited its benefits without altering the natural phenomenon. Only in the last hundred years has man developed the tools to change the environment of Louisiana at a landscape level. On some planes we may be progress poor.

Ecotone

The term "ecotone" can be defined as a transition area between two adjacent ecological communities. It usually has some common characteristics of each bordering community and often contains species not found in either of the two. Ecotones exist at different scales. It may be the edge of your back yard where it butts up against a bayou or patch of woods. It can be a 20-mile-wide strip that separates the eastern front of the Rocky Mountains from the Great Plains or the northern evergreen forests from the tundra.

In north Louisiana a distinct ecotone generally follows the path of the Ouachita River. Historically, this ecotone separated the very different bottomland hardwood forests east of the river from the mixed upland hardwood/pine forests west of the river. The actual ecotone is at the western bank of the river, in some cases, where the hills come down to the water's edge. In other areas where the river has meandered away from the hills, the ecotone is farther from the river. It can be thought of as that area where the red clay hills drop off into the swampy bottomlands. Other ecotones in Louisiana included those between the freshwater marshes and prairies of southwest Louisiana, and the longleaf pine and upland hardwood forests in the Florida parishes.

One important aspect of ecotones is their value as wildlife habitat. Generally speaking, the more different types of plants found in an area, the more kinds of animals live there also. Since ecotones have plants common to both adjacent areas, many animals have an easier time making a living there. In the north Louisiana example, squirrels have the opportunity to feed on acorns from oaks that only grow in the swamps and also on those types that grow on the drier hill soils. Several species of birds thrive in ecotones along forest edges. Mockingbirds and indigo buntings live on the edge of woodlands but are rarely found deep in a forest.

Humans also took advantage of the ecotone along the Ouachita River. The abundant and diverse natural resources of the hill and swamp forests found in close proximity may be one of the reasons why the North American mound-building culture began along this dividing ecotone. As a practical concept important to their daily survival, early Native Americans could have defined the word "ecotone" better than we can today.

Habitat Fragmentation

In recent years the term "habitat fragmentation" has become a common buzzword and concern among conservationists. Habitat fragmentation occurs when a large natural area is altered so that only scattered fragments of the original remain. Examples in Louisiana include converting parts of a bottomland hardwood forest to farmland or cutting much of an oak/hickory stand in the hills and changing it to a pine plantation. Thousands of acres in all of the state's ecosystems have undergone habitat fragmentation from oil- and gas-related energy development. On a smaller scale, habitat fragmentation in Louisiana is a result of residential, commercial, and industrial development.

In addition to outright destruction of habitat, fragmentation is disruptive for animals left in the intact areas. Native birds are especially vulnerable. When a forest is fragmented, the amount of edge around the remaining forest increases relative to the overall size of the tract. The impact is amplified if clearing occurs in the interior of a forest. An increase in edge effect is harmful in at least four ways for many songbirds that are adapted to live in the interior of large tracts. First, the rate of nest predation is higher. Several species of nest predators, such as raccoons and crows, are more abundant along forest edges. Second, the incidence of brood parasitism is higher. Brown-headed cowbirds, also more abundant along forest edges, lay their eggs in the nests of other birds and are unwittingly raised by their foster parents. Young cowbirds are aggressive and usually displace the young songbirds in the nest. Third, interior forest birds have more competition for good nesting sites with those species that are adapted to the edge. Fourth, interior-forest songbirds often have lower overall nesting success if they are forced to nest near an edge. One study revealed that only 18 percent of nests within 100 meters of a forest edge were successful, while 70 percent of nests greater than 200 meters from an edge were successful. Some birds are so area sensitive that they can maintain sustainable populations only in large forests. It is estimated that Swainson's warblers need a forest block of at least 10,000 acres, cerulean warblers need 20,000 acres, and swallow-tailed kites need 100,000 acres. The only 100,000-acre forest in Louisiana, the vast Atchafalaya Swamp, harbors the only breeding population of swallow-tailed kites in the state.

The need for large, unbroken tracts of forest to support animals such as the Louisiana black bear is more obvious. Their big home ranges are necessary to find enough food and for successful reproduction. The forest must be capable of supporting enough bears to overcome genetic inbreeding problems associated with small populations. Large predators, such as cougars and red wolves, disappeared from Louisiana in part because of forest fragmentation.

The concept of fragmentation and its consequences demonstrate the need for a deeper awareness of our remaining natural habitats. Knowing the total amount of forested land in Louisiana is not enough. We need to consider the size of individual forest blocks and their location in respect to other blocks. With this information we can make better management decisions and concentrate limited resources to protect the most vital tracts. We can route new roads and rights-of-way away from forest interiors. We can join and develop corridors between tracts with reforestation projects. Perhaps most important, we will be able to accurately predict the consequences of additional fragmentation in our remaining natural areas.

Hibernation

A while back I read in a popular travel guide a specific warning to tourists traveling in Central America. This advisory, among others, stated: "Don't place hands in snake holes." Go figure. Apparently this is a problem, and we'll get back to the anecdote at the end of this story.

Wild animals adapt to the cooler temperatures of autumn and winter in various ways. Many types of mammals grow thicker coats of insulating hair. Migratory birds pack up and fly off to warmer climes. Some species, though, undergo drastic physiological changes in order to survive. The often-misunderstood term "hibernation" is used to describe these adaptive states.

Hibernation is a condition of inactivity usually stimulated by low temperatures and characterized by lower metabolic rate, lower body temperature, and slower breathing. Various degrees of hibernation occur. True hibernation is defined by a deep torpor. A true hibernator cannot be quickly aroused and appears dead. Their body temperature approaches the outside temperature, often near freezing. Some species of bats, rodents, and snakes are true hibernators.

Other animals undergo less extreme metabolic changes. Species with very high metabolic rates, such as hummingbirds and shrews, often go into a short-term torpor during cold nights in order to save energy. Two or three hours of morning sun may be necessary to warm them back to a functional state.

Another condition similar to true hibernation is estivation. Animals that estivate are dormant during hot summer periods to avoid heat or lack of food and water. Some tortoises, salamanders, and lungfish use this strategy.

One common misconception in Louisiana involves the idea that bears hibernate. They do not. Their dormant activity can be more properly termed "denning." Bears accumulate body fat during the summer and autumn. They enter dens, usually large hollow trees or brush piles in our area, and spend a lot of time sleeping. Bears are quite capable of surviving Louisiana winter temperatures outside, and their inactive periods seem to be an adaptation to a season of scarce food supplies. It's just easier to pig out during the good times and mostly sleep away the winter. Scientists quickly learned that bears are not classic hibernators when they attempted to insert thermometers into the denning bruins. Indeed, they remain alert enough to neutralize any such insults. So, in case you suffer from "instinct deficient syndrome," the take-home messages are: don't try to take the temperature of a denning bear; like some reptiles, they are not true hibernators, and accordingly, don't place your hands in snake holes.

Marking Bayou Time

February 16—red swamp maples blooming
February 28—chorus frogs calling
March 3—dogwood buds swelling
March 19—gray tree frogs calling
March 20—first dragonflies spotted, white violets blooming
March 24—first wood thrush flutes
March 26—Christmas ferns unfurling
March 28—first luna moth appears
April 1—red buckeyes blooming, ruby-throated
          hummingbirds return
April 10—first bullfrog and screech owls calling, fireflies
          appear
April 18—chuck will's widow calling
April 20—big water moccasins sunning in Rocky Branch
May 16—box turtles and skinks mating
May 25—tiger swallowtail on butterflyweed

May 28—baby chickadees & tufted titmice at feeder
May 30—narrow-mouth toad buzzing

The study of the cyclic events of nature in response to seasonal and climatic changes is called phenology. In simpler terms, it is just "paying attention" to outdoor happenings. Today the rhythm of the natural world vanishes inside air-conditioned cocoons with blaring televisions and video games for most people. That lost is a source of inspiration, motivation, and salvation from the stresses of the twenty-first century.

In the days when sitting on the front porch was a primary form of entertainment, natural happenings were as familiar to folks as their friends and family. They watched the size of oak leaves and knew when it was time to plant corn to avoid a frost kill. They watched the phases of the moon and knew when to most easily dig a posthole. They watched the sky, not the weather channel, to predict the weather. They could count cricket chirps and calculate temperature.

The natural world operates in an ordered flux. Everything happens on a schedule. When small flying insects hatch, predatory dragonflies emerge from their nymph stage and begin feeding flights. Frogs metamorphose from tadpoles when their food becomes available. Hummingbirds return when the red buckeye blooms, and the tiger swallowtail arrives in time for butterflyweed nectar. There is a script for nature's play.

Find time in your busy life to step outside, sit quietly, and pay attention. Become a silent witness to the rhythms of natural life around your home. There's a lifetime guarantee of amazement.

Pollen

For the past several days I have been witness to an orgy, the likes of which would have titillated the goddess Aphrodite had she understood it. This reproductive frenzy is botanical in nature and is re-

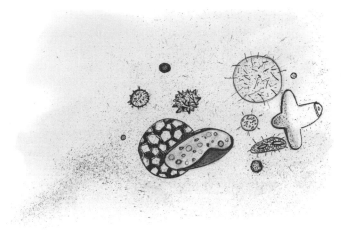

Pollen

vealed by the clouds of breeze-borne pollen that blankets every horizontal surface.

Pollen is a powder-like substance containing the male sexual cells of flowering plants. Even in a small meadow or patch of woods, billions and billions of pollen grains are released for the sole purpose of procreation. For this to occur, pollen must come in contact with the female part of a flower of the same species. Barring compatibility problems, fertilization occurs and seeds are produced. Life goes on. In some species, the plant uses the pollen from its own flowers to fertilize itself. Other types must be cross-pollinated; that is, pollen from the flower of one plant must be transferred to that of another plant of the same species for fertilization to be successful.

Depending on the type of plant, pollen is usually transported either by the wind or by insects. Plants with large or showy flowers that attract insects often depend on them for pollination. Species with small, plain flowers produce light, dry pollen grains ideal for wind transport. At least a quarter of a million plants use the wind for pollination.

Herein lies the source of misery for untold numbers of allergy sufferers. The random flight of airborne pollen often ends in warm, moist human throats and noses, triggering an allergic reaction known as hay fever. The chemical composition of pollen determines whether it is likely to cause problems. For example, pine trees in Louisiana are among the most prolific producers of pollen, yet the chemical makeup appears to make it less allergenic than most. Local trees that do produce allergenic pollen include elm, ash, oak, hickory, and pecan. Among grasses, Johnson grass and Bermuda grass are culprits. Those plants commonly referred to as weeds beget most allergenic pollen. Ragweed is infamous, but others of importance are pigweed, lamb's quarters, and plantain. Keep in mind that the wind commonly carries pollen for many miles so chopping down the pecan tree in the front yard is a waste of time, effort, and pecans.

Pollen from different plants occurs at different times of the year in the various parts of the country. In bayou country, most trees produce pollen from January through March. Grasses reproduce from April through November, and weeds peak in the summer and early autumn. Only December is relatively pollen free.

The local media often report daily pollen counts. This count is a measure of the concentration of pollen in the air at a specific time. Counts are usually highest on warm, dry mornings and lowest during chilly, wet periods. At best, counts are only indicators of the best time to stay indoors. As a diversion from the effects of a spontaneous sneezing fit, allergy sufferers should consider the sensuous nature of the pollen that caused it.

Pond Spoor

Deer, gray squirrel and gray fox, possum, raccoon, armadillo all come to the drying pond now, leaving their spoor in the encircling halo of mud. What a difference between equinoxes. Six feet deep in the spring, the pond reflects the harvest moon from a surface barely 18 inches above the muck bottom. Daily evaporation sucks away at the pond's diameter. Except for the squirrels, most of the mammals come at night or at the crepuscular times in between. Deer, some heavy in their splayfooted tracks, others with hooves of the year barely larger than a nickel, wade into the tepid water to drink. The armadillo trudges like a tank plowing a trail into the shallows that smears his three-toed prints with a dragging tail. He is more amphibious than most people know. Possum's mark is the imprint of her hind foot with a toe that appears as an opposable thumb. Her kind has ambled about the older parts of bayou country since their northward trek from South America three million years ago. She is resilient. The raccoons come to eat as well as drink. With a refined sense of touch that processes stimuli in dark places, receptor-laden paws probe crevices and burrows for the delicacy of a molting crawfish. Like a dealing card shark, they look away from busy hands while plying their trade. Squirrels come to drink in a stealth mode, creeping in to stretch out full length on soft bellies before quenching the thirst with rapid lapping. If only they could control the demon-possessed tail, their subterfuge would be complete. The gray fox would roll up his britches legs if he could. His tracks never enter the water but meander along the shore behind his nose. Maybe he doesn't even drink but comes to meddle in the business of those in Order Rodentia. On a landscape scale the pond is a tiny dot on the edge of a large swamp. For most of the year it is of little consequence in the lives of these animals as the nearby wetlands pulse with seasonal overflows. Now, though, the bowl contains the essence of their existence.

Wildlife Diseases

Medicare, Medicaid, cancer, and heart disease are a few examples of human health issues that seem to constantly make headlines. But even for many of us who are outdoor enthusiasts, the health of Louisiana's native wildlife populations is rarely contemplated. Diseases of wildlife are not new. In fact, they have been recognized for centuries. More than two thousand years ago, references to diseases in wildlife were recorded in the Bible, as well as in the writings of Homer and Aristotle. As a science, though, this subject has only developed in the past fifty years.

Diseases in wildlife are often similar to those in humans, the causative agents usually being bacteria, viruses, or parasites. In many instances these diseases are a natural part of an ecosystem and serve to keep populations in balance with their habitats, just as predators sometimes do. Diseases of deer have been well studied, and herd health checks are routinely conducted in Louisiana. One standard test, which determines the number of stomach parasites, yields results that correlate with the general health of a herd: high parasite count = poor herd health, usually as a result of too many deer in a given area, or low parasite count = good health and a herd in balance with its habitat.

Mortality associated with wildlife diseases frequently goes unnoticed because it doesn't often occur on a large scale in a short period of time, and scavengers quickly recycle the evidence. Exceptions exist, particularly with reference to waterfowl. In the winter of 1965–66, an estimated seventy thousand birds died of avian cholera in north-central California. Even larger outbreaks involving diving ducks have occurred in the Chesapeake Bay. Birds are also subject to botulism, tuberculosis, salmonellosis, avian pox, and sarcocystis. Local wild mammals must contend with hemorrhagic disease, anaplasmosis, leptospirosis, and rabies, among others.

The occasions when wildlife diseases are not natural cycling mechanisms in Louisiana ecosystems can always be traced to humans. Toxic diseases, such as lead, mercury, and pesticide poisoning, result from human insensitivity to wildlife. Humans have introduced alien diseases into susceptible native wildlife via pets, domestic animals, and exotic species. More than any other, these diseases pose serious threats to local wildlife and serve to remind us of the vulnerability of wild animals to disease, both natural and unnatural.

Umwelt

We humans describe the world that we see around us as our environment. Our perception of this world is unique in that none of the other living creatures on the planet share our experience. Each species has its own awareness of what we call the environment. The term for what an animal perceives is "Umwelt," or self-world. There can be as many different Umwelten in a particular environment as there are kinds of animals.

The Umwelt for any creature is dependent upon the types of sensory receptors that it possesses and its capability to process stimuli of those receptors. Our awareness is based on the five senses of sight, smell, touch, hearing, and taste, each with its own limits. Various animals have these senses and often more that function within different parameters. For example, humans have about 4 square centimeters of membrane that detects odors. Dogs have 150 square centimeters of this organ to find the perfect stench to roll in. Humans can hear sound frequencies between 20 and 20,000 hertz. Elephants can hear in the ultra-low range of 1 hertz, and some moths can hear frequencies of 240,000 hertz. Even Chef Paul Prudhomme has only 10,000 taste buds, but the catfish in his famous dishes have 100,000. A burrowing mole has 6 times more touch receptors in his nose than in the human hand. As for sight, the normal vision for people is 20/20. A hawk's vision is equivalent to 20/5.

And then there are those animal senses that we can barely imagine. Whales and bats use echolocation to navigate. Rattlesnakes use infrared vision to hunt warm-blooded prey in the dark of night. Fish use electric fields to monitor their surroundings. Bees, sea turtles, migratory birds, and tuna can detect the magnetic fields of the Earth and orient themselves with this mysterious process. It is easy to forget that the wildlife around us view reflections of southern sunsets on our bayous from a different perspective, one defined by their own Umwelt.

Transpiration

All around us, especially in the hottest dog days of summer, a silent sucking sound permeates the oppressive humidity. It is the water in plants being pulled from the roots to evaporate through microscopic openings called stomata on the bottom of leaves. The process is transpiration, and it is most dramatic when considered on the scale of trees. A large oak in your yard can release hundreds of gallons of water on a hot, dry day, up to 40,000 gallons in a year. Plants release about 10 percent of all the moisture in our atmosphere; the rest comes from ocean evaporation. The amount of water lost by a plant depends on its size, temperature, humidity, wind speed, and soil moisture content. Transpiration serves three critical roles in plants: the water movement provides an avenue of transport for minerals and food throughout the plant; it cools the plant (and, incidentally, humans, as 80 percent of the cooling effect of a shade tree is from the evaporative cooling of transpiration); and it maintains turgor pressure in cells, which allows plant parts to remain firm and upright. When the amount of moisture in the soil fails to keep up with the rate of transpiration, loss of turgor pressure occurs and the stomata close. Transpiration plummets and the plant wilts. Simply put, when your dogwood looks hang-dogged it's because the soil under it is too dry to support normal transpiration, and the tree struggles to adapt by entering a survival mode. On a Louisiana summer day it takes only a bit of imagination to hear the tons of water slowly moving up through roots and tall trunks and out the tiny leaf openings to collectively coalesce into a small cumulus cloud, upon which we base our hopes of a cooling shower.

Cold Weather

A recent spell of severe winter weather disrupted the daily lives of many people in Louisiana. Technological advancements in the last hundred years, however, minimized the impacts to a short period of inconvenience for most. Consider the differences now and during the Civil War as described by a Confederate soldier. He wrote, "The last days of the year were rainy and disagreeable, and on the last night of 1863, the rain ceased, the wind blew almost a hurricane, turning the air most bitterly cold . . . There is not a man of the 18th who forgets the march on that New Year's day, up the river and thru Monroe. The wind came directly in their faces . . . freezing so rapidly that the earth was frozen hard enough to bear up our heavy wagons . . . Many of the men had their fingers and feet frozen . . . many . . . were barefooted, some were without blankets, and others almost destitute of clothing . . . On the march to Bartholomew Bayou, the blood from the feet of the men was frequently to be seen upon the frozen ground." This passage tends to put our recent weather-related privations in perspective. For better or worse, humans were much more in tune with the environment in those days.

For native plants and wildlife, though, little has changed. Cyclic weather extremes configured our natural ecosystems and continue to do so today. We can even predict with a degree of certainty some results of hard freezes based on past observations of similar events. For instance, south Arkansas tends to be about as far north as alligator populations can consistently thrive. Cold weather is a limiting factor, and many alligators in that area die in severe cold spells. Woodcocks are quail-size birds that spend the winter here and are adapted to feed on earthworms with their long, probing bills. Large numbers have been known to starve when the ground freezes hard for several days. Waterfowl wintering in Louisiana stay busy trying to build up fat reserves for the long flight back north to their breeding grounds. A stressful weather event can cause some, especially females, to arrive in less than optimum breeding condition, and thus have a negative impact on the number of young produced in the spring. In a recent year, oak trees produced a larger than average crop of acorns, but few germinated because a severe freeze killed the embryos inside. In contrast, some native species benefit when exotic invaders, including fire ants, nutria, water hyacinth, and Chinese tallow, are suppressed by cold weather.

Generally, isolated weather events have little long-term impact on native plant and wildlife populations. They contribute to nature's complex system of checks and balances, which has evolved over thousands of years. The remarkable resiliency of our ecosystems is challenged only when human technology overloads the system.

Hot Weather

Always during the play of August days in Louisiana when the temperature taps the century mark, the sun breaks first on the horizon like a glowing, fertile egg yolk. If there has been no recent respite in the Earth's natural fever, all life awakens with a bunker-style strategy of survival. Many species of deciduous plants begin the morning in a wilt instead of waiting for the afternoon. With roots unable to replenish the scarce moisture as fast as evaporation sucks it from leaves, even the dogwoods are hang-dogged. Creatures of the diurnal world stir early and late or not at all. Mammals, haired for temperate climes, lay low. Birds forage at dawn in a brief flit of activity before seeking shade, where young-of-the-year cardinals and titmice will pant like puppies by noon. Of their kin only the enigmatic vulture seeks the sun and death in orbits barely visible in the glare. The cold-blooded creatures are not. This is their time but only if careful. Equipped with thermometers but lacking a central heating and cooling system, skinks and anoles dare not tarry on a hot rock lest they stew in their own juices. A rustle of dry leaves reveals their quest for those without backbones—sugar ants and the early instars of grasshoppers. Meanwhile, cicadas drone the song of this, their only season. All else is silent and waits.

The drama of this scene may last for days although usually within a fortnight the spontaneous forecast of a skulking rain-crow becomes more than a haunting lament in the boughs of a white oak. On a late afternoon the first signs are just sensed as low-frequency waves shove through the humidity and the barometer marks a change. Antennae and whiskers twitch, lenticels respond. Then, simultaneously, a cooling zephyr and thunder arrive from the southwest. Soon the black mountain appears, trees thrash bitterly among themselves while shedding the twig-girdled clumps of brown leaves favored by vireos, and great dollops of rain close the act for a while.

Homeostasis

Homeostasis is an important ecological concept that impacts our daily lives in many ways. Homeostasis means "resistance to change" or "staying the same." Operating at many levels, it is the process by which a cell, an organism, an ecosystem, or the entire planet maintains constant internal conditions in the face of a varying external environment.

An example of homeostasis at the cellular level is the phenomenon in which division of a group of cells stops when they become so numerous as to touch each other. Biologists think that at this time a chemical that halts further division is passed from cell to cell, thus the system resists change and stays the same. Cancer cells appear to have overcome the chemical resistance and continue to divide even after cells touch.

At the organism level, homeostasis has survival value by allowing animals to adapt to changing components of the environment, such as solar radiation, temperature, and food supplies. When it's cold on the bayou, we shiver and warm up a bit. When broad-headed skinks are hot, they seek the shady side of a cypress tree. Other strategies for maintaining heat balance include migration, hibernation, and growing thicker fur or darkened skin. The sensations of hunger and thirst are also homeostatic mechanisms that help an organism maintain optimum levels of energy, water, and nutrients.

Homeostasis can also regulate the size of populations, as shown in the relationship between predatory animals and their prey. When snowshoe rabbits become abundant in Canada, so do their main predators, the lynx—until predation diminishes the supply of rabbits, which causes a corresponding decline in lynx populations.

When rabbit numbers build again, the cycle is repeated, and the general fluctuation remains stable and resistant to change.

At the global level some think that all living matter on Earth functions as a vast organism that actively modifies the planet to produce a suitable environment, thus the entire planet maintains homeostasis. While this theory is debated, some simple global mechanisms are generally accepted. For example, when carbon dioxide levels in the atmosphere rise, plants are able to grow better and remove more carbon dioxide from the atmosphere.

Finally, large organizations, industrial firms, and social systems are sometimes referred to as homeostatic because they oppose change with every means at their disposal. That's out of my field, but I do know of a person or two whom I formerly called hardheaded but will now label as homeostatic.

Soil

If you live in Monroe, much of West Monroe, Sterlington, Lake Providence, or Tallulah, your dirt is young—10,000 years old or younger. These are alluvial deposits, sediments laid down in the floodplains of flowing rivers and streams. If you live on the Macon Ridge in Oak Grove, Delhi, Rayville, or Winnsboro, your soil is older, perhaps up to 20,000 years old, and has a cap of loess—a windblown deposit. Bastrop is on a separate island of this soil type. If you live in the red clay hills of West Monroe, Ruston, Farmerville, or Bernice, your soil is very old and formed in the Eocene epoch 40 to 55 million years ago. A very small area southwest of Cheniere Lake in Ouachita Parish even has soil more than 60 million years old. Soil and rocks are aged by various techniques: radiocarbon dating, the degree of weathering, and the degree of soil development. It is not an exact science.

All of Louisiana has been completely covered by shallow seas at some point in geologic history. Likewise, much of the land at lower elevations has been inundated by flooding rivers, bayous, and streams. All of our surface soils (except loess soils) were laid down when the land was flooded as sediments such as clays settled out of the muddy waters. The older soils in the hill country were formed when sediments fell out of shallow seas; sediments of the flooding Mississippi River and its tributaries formed the younger soils in the delta.

Soils dictate fertility, moisture regimes, and the type of vegetation that will grow. Pines and upland hardwoods grow in the hill country; bottomland hardwoods in the delta. A very unusual soil type near Columbia in Caldwell Parish supports plants found nowhere else in the state. Soil sustains billions of organisms below its surface in the form of insects, worms, and bacteria. It provides a medium in which plants can grow and nourishes them with nutrients. Soil filters and purifies rainfall to recharge the aquifers that supply our drinking water. It is essential for life on Earth and will eventually recycle our remains, whether or not we have poetic tendencies.

Storms I

When severe storms occur in north Louisiana, their impact is most often observed on trees. Tornados and other windstorms uproot trees and break branches. Hail and ice storms strip leaves and limbs. The results can be dramatic. In the coastal areas of Louisiana, storms have different effects. A good example is the recent tropical storm that struck the vast marshlands around the mouth of the Atchafalaya River. Trees are scarce on this active delta, and the topography changes continuously as currents and sediments make maps obsolete in a year's time. Ponds and lakes fill, new channels are plowed, and ephemeral lands emerge first as mudflats and later as bona fide marshes. Vegetation consists of annual and perennial aquatic plants rarely more than waist high.

When the high winds of Tropical Storm Francis buffeted this area, the physical forces destroyed the leaves, making the marshes appear to have been scraped with a giant cheese shredder. After the initial assault, the wind-borne sea invaded with high tides 2 feet higher than normal. Saltwater sickened and poisoned the less tolerant vegetation. The scene after the storm left little doubt a significant natural event had just occurred.

The storm also started a chain reaction that affected the entire ecosystem. Dead vegetation began to decay immediately in the humid, subtropical climate. An odor of brewing microorganisms was omnipresent. Water rushed out of the marshes into the bayous and created standing waves a foot high where currents collided in the relentless tolling of gravity. Uncountable numbers of small, herbivorous fish were sucked into eddies and held hostage for the inevitable arrival of avian predators. They came by the thousands, different species of gulls and terns, even the endangered least tern, to plunge into the masses of silver scales.

On the second day, life below the surface became even more difficult. Rotting vegetation left the receding water short in one critical element—oxygen. The more susceptible bass, long-ear sunfish, and gizzard shad began piping at the surface. By the third morning, their bloated carcasses were floating to the top. A second assault of birds, these from a different taxonomic order, arrived as if on cue. Long-billed and long-legged, little blue herons, great blue herons, Louisiana herons, green herons, yellow-crowned night herons, great egrets, and snowy egrets, each in their specialty, reaped the results of a literal windfall.

This image of death and destruction does not portray catastrophe in the natural world. Storm events are common and even necessary in some instances to perpetuate life. The spring following a north Louisiana ice storm would still bear evidence of the event in broken limbs and shattered tree trunks. The spring following Tropical Storm Francis in the Atchafalaya marshes appeared innocent of the recent havoc. The fast-growing vegetation recovered, and fish populations rebounded. The flimsy, stick nests of wading birds and the mound nests of alligators were full of young—the results of their parents' good health through the winter as a consequence of abundant food after the storm. One might conclude that there is indeed order in natural chaos.

Storms II

From most people's point of view, the recent storms that brought strong winds were destructive and anything but beneficial. If a forest were capable of a perspective, it would be that such storms are critical to their survival. Most of the forested lands in Louisiana once had some species of oak as a major component. The exception was the region blanketed with longleaf pine. Alluvial bottomlands were covered with several types of oaks and other hardwoods. The hill country grew different types of oaks mixed with pines. These forests were mosaics of trees of different ages. Large stands of even-aged trees were rare.

Oaks have growth habits that foresters refer to as "shade intolerant," meaning that young oaks cannot grow in the deep shade of other trees. It is not uncommon to see hundreds of oak seedlings under a parent tree in the forest. Most, however, will never grow taller unless the big tree falls and sunlight reaches the seedlings. During the infrequent storm events in which strong winds occur, the most susceptible trees to windthrow were the largest and oldest individuals. If the adage that asserts "the fittest survive" is considered, one might conclude that the big trees that were toppled were less fit than the younger, stronger trees that withstood the winds. The opposite is true. The fact that the big, old trees reached such size and age in the first place proves they were more vigorous, disease resistant, and stronger than their neighbors. They were the fittest, and they survived through their offspring that were released from the bonds of shade when the old trees fell.

Most forests in the state today are even-aged. They originated when earlier forests were cut down, and the new trees all began growing at the same time. Trees in these forests will be cut long before they reach the size and age of their ancestors. A windstorm in a modern forest *will* fell the weakest trees and perpetuate the survival of their correspondingly weak offspring below. Over time, the general health of the forest will decline. Human manipulation of the environment on a large scale can even reverse the natural role of a windstorm.

Storms III

This story was written soon after the infamous hurricanes of 2005.

Hurricanes Katrina and Rita plowed up the Louisiana Gulf coast, leaving a wake of human suffering and tragedy now familiar to people around the world. Less well known are the impacts on fish and wildlife of the region. We will be still be learning these effects in years to come.

Although scientific studies have not yet been conducted, anecdotal information is plentiful. Marshes are littered with dead alligators. Even this apex predator of the wetlands can drown in a raging storm surge. Extensive, widescale fish kills have been reported. A combination of saltwater contamination and low oxygen levels resulting from decaying vegetation proved untenable for freshwater species. Several porpoises were washed far inland and stranded, as was a sea turtle that was eventually rescued. A threatened Louisiana black bear was forced from flooded swamps to the high ground of a highway and struck by a vehicle. As an adult female in a population of only a hundred, her loss is significant. On a grander scale, vegetation that provides food and shelter for many types of animals on hundreds of thousands of acres of marshlands was destroyed by the surge of saltwater. The fate of those animals that survived the initial storm is unknown. Soon, millions of waterfowl that normally migrate to coastal wetlands will arrive to find a habitat devoid of critical food plants. Can they adapt by moving elsewhere?

Damage to manmade structures poses unique problems. An estimated 7 million gallons of oil products spilled into the environment. Tons of debris, including contaminated fuel tanks, litter the marshes. Levees built specifically to benefit fish and wildlife were destroyed. A visitor center at Sabine National Wildlife Refuge that taught the value of wetlands to countless people was reduced to foundation pilings—a loss that is immeasurable.

Hurricanes are natural events and in some cases are actually beneficial. Invasive water hyacinths and nutria were killed, and some dunes were replenished with sand. On the whole, though, the environmental ledger recorded a huge deficit from the storms. The Louisiana coastline, already on life support for a host of reasons, most manmade, just could not absorb the events that many now consider warnings of worse to come.

Tree Rings

One day a forester walked into an unfamiliar patch of woods. She chose a medium-sized white oak tree as a subject, collected a sample, and returned to the lab. In a couple of days she drafted the following history of the tree.

Tree Rings

"The white oak sprouted from an acorn in the year 1885. For the first twelve years of its life, the tree grew slowly as a result of being shaded from sunlight by large trees nearby. In 1897, the closest large tree, perhaps a parent of the sapling, blew over in a spring storm and allowed sunlight to reach the young tree. It grew rapidly for the next seventeen years. Then, beginning in 1914, a severe drought slowed growth for three years until favorable precipitation returned to the area. For the following forty-two years growth was normal. In the fall of 1959, a fire raged through the forest, severely injuring the white oak. It survived because of its thick bark but barely grew at all for eight years as a pathogenic fungus attacked through the fire scar. In 1963, a neighboring tree to the east fell hard against the white oak, causing a permanent lean to the west at a thirty-degree angle. At the time of this sampling in 2010 the tree is 125 years old."

Although this story is hypothetical, it is typical of many life histories of trees that are determined through the science of dendrochronology. In temperate climates trees form growth rings in response to changing seasonal conditions. Each ring has two parts: a wide, light part called the early wood, and a narrow, dark part known as late wood. The early wood is formed during the wet, spring growing season. The late wood forms during the drier transition period from summer to autumn and winter when growth slows. By studying the rings, information about the climate over a period of time and evidence of environmental disturbances, such as fires and floods, can be learned. The shape and width of the annual rings differ from year to year because of varying growth conditions. A ring formed in a wet favorable growing season may be very wide, while rings formed when the tree is stressed will be much narrower. Fire scars and insect damage are often visible in the rings.

To study a tree's growth rings without harming the tree, scientists use a technique called coring. A hollow, tubular instrument known as an increment borer is used to drill into the center of a tree trunk and extract a narrow cylinder of wood. Growth rings on this core sample appear as lines that can be counted, measured, and studied for abnormalities. Similar observations can also be made by examining the stump of a freshly cut tree or the end of a sawn log.

By studying tree rings in very long-lived species, such as bald-cypress, and overlapping data from living trees with that obtained from old fallen logs, climate, fire, and flood histories can be developed for an area that go back more than two thousand years. This information helps us understand how forests change over time, and in some cases can be used to predict the frequency of future fires and floods.

So as you enjoy the shade tree in your back yard, remember that history and lessons for the future are being recorded season by season just under the bark.

Migration I

The ridgeline of the roof of my house in the woods is oriented north-south. In October this can be determined without a compass, without watching the sunrise or sunset, and certainly without checking to see which side of the white oaks that moss is growing on. From the back deck one has only to watch the sky through the hole in the canopy above the house. If the timing is right, avian compasses appear in the form of small groups of songbirds. The individuals are bunched tightly together; their flight is hurried and gives the impression of a communal determination. They fly right to left straight down the roof ridgeline and disappear in seconds. They are a part of one of the most mysterious acts of nature—bird migration. Even after years of scientific study the exact methods birds use to navigate across continents and oceans in biannual journeys are poorly understood. Some are known to use the sun as a reference point, but that assumes that birds also must have an internal clock in order to know where the sun should be at a given time of day. And what about those species that migrate at night? Tests have shown that some orient by the stars, a conclusion made with the caveat that these birds must also possess a knowledge of the night sky—a type of preprogrammed star map for each night at every location on their migration route spring and autumn. Then there is the issue of how birds orient when the sun and stars are obscured by clouds, which they seem to do with ease. It turns out that the same iron atoms in the metal roof of my house are found deep inside the brains of some birds, and they are sensitive to the Earth's magnetic pole, thus serving as a built-in compass. Watching these birds hurry on their way, it is comforting to me that every revelation hatches a dozen new questions.

Migration II

If someone were to invite me to go birdwatching along the bayou in the middle of the night, I would suspect shenanigans of the mythical snipe hunt. However, such might not be the case, as there are opportunities to birdwatch at night in Louisiana. September and October are some of the best times.

During this period, fall migration is in full swing, and it is common to see flocks of hawks, vultures, and pelicans riding the afternoon thermals. These birds that soar and glide on fixed wings usually migrate during the day to take advantage of vertical updrafts generated when the sun heats the Earth. On the other hand, birds that employ wing-flapping flight generally migrate at night. Most songbirds, shorebirds, and many waterfowl travel under the cover of darkness.

For years, nocturnal migration was thought to be an adaptation to allow birds to avoid predators or to effectively navigate using the stars. Neither of these theories explains the behavior adequately for all species. For instance, many large birds that often migrate at night, such as herons, ducks, and geese, are rarely at risk from avian predators, and flights of various species on cloudy, starless nights are common.

Recent theories suggest that nocturnal migration is based on atmospheric conditions. During the day, the atmosphere is less stable as the sun heats the Earth. Airplane passengers frequently experience the bumps and jolts of turbulence. Imagine the effect on a 2-ounce songbird. After dark, thermal activity ceases, the air smoothes, and small birds can travel more efficiently. Air temperature also affects flight. Flapping wings generate body heat, and birds must be careful not to overheat. Flying at night when the air temperature is cooler is an advantage. Radar studies along the coast convey the scope of nocturnal migration in Louisiana. During peak migration, up to fifty thousand songbirds per hour pass through each linear mile scanned by radar.

To experience this timeless phenomenon, choose a bright moonlit autumn night. A half to full moon is best. Focus a pair of high-powered binoculars, or better yet a telescope, on the disc of the moon, and watch for dots that move across the moon's face. Sometimes you can even hear faint birdcalls—messages from tiny travelers responding to ancient urges.

Swamp Snow

Swamp snows don't come often to Louisiana. Only during a rare conjugal visit of otherwise estranged weather gods, warm wet air from the Gulf overrides a lingering cold front to produce moisture that morphs into hexagonal crystals. If the snow seeds are sufficiently fertile and the humidity high, flakes the size of dimes, nickels, or even the wings of bride moths float into the winter world of baldcypress trees, Spanish moss, and squealer ducks. Almost always the temperature is marginal, the apparition fleeting as a persistent sun sweeps clean the spell in a cruel shower of snowmelt. It is best to visit a snowed swamp

Swamp Snow

soon while the sky is still leaden, to eschew the garish glare in favor of shadowless hues, subtle and natural.

Tree bark and slough water provide contrast for the whiteness. Willow oaks have coarse, dark-roast coffee bark; the skin of cypress is furrowed russet. All things botanical, apart from the vertical, wear ermine mantles, especially the logs on their journeys back to earth. Members of the wetland arboretum appear to doze and transpire slowly under their insulating blankets. The water is translucent black, and cold as liquid water can be. It is swamp blood, sustained now with snowflakes as well as raindrops. As molecules flowing across the gills of widow skimmer dragonfly larvae, they are not discernible. Pumped through xylem 80 feet up to the highest twigs of an overcup oak in order to nurture an acorn, it matters not what form they entered the swamp. Here contrast is absorbed.

Louisiana swamp snows bear other gifts in the shape of anomalies. Orb spiders in their webs snare snowflakes instead of mosquitoes. In the frigid water wood ducks preen, cavort, and squeal in anthropomorphic displays of delight. Emerald mosses go about their subtropical business of procreation, and fish crows fly over without ever uttering a word. They know that all traces of the day's conjuration will vanish on the morrow.

CONFLUENCE
A Flowing Together of Two or More Streams of
Life in a Physical Place

4

ENCOUNTER

A Meeting, Especially One That Is Unplanned, Unexpected, or of Note

On night of 5 June, I motored to northeast spillway of Lacassine Pool in Whaler. Carried pirogue to photograph in Pool. Photographed alligators, raccoon, pig frogs, lotus blooms, green snake and marsh. Ran aground [off Lacassine Bayou] in Whaler on return trip and had to paddle and pole pirogue 5 miles home. Arrived 2AM.

—KO Field Diary, 6 June 1981—Lacassine NWR

Went w/Keith Weaver to check bear snares on Deltic property north of Hwy. 80. Caught 229 lb. male on Wade Bayou Unit—drugged him, put in ear tags, radio collared him, tattooed lip, measured & weighed him, pulled a pre-molar, took blood & tissue samples—sewed him up, gave him antibiotics—he was beautiful in great shape; My 1st Louisiana bear!

—KO Field Diary, 1 June 1988—Tensas River NWR

Tributary Affair

Shortly after New Year's Day in 1913, the steamboat *Gopher* pulled hard to port side and entered the mouth of Bayou D'Arbonne from the Ouachita River above Monroe, Louisiana. She was on a special chartered trip, and the passengers had no interest in the bustling packet trade that usually paid the captain's bills. One of them was a world traveler and a man of letters from the Philadelphia Academy of Natural Science. His companions knew little of letters of any kind and wore hard labor calluses with their overalls.

The bayou was half a century away from knowing the biodiversity-choking effects of locks and dams. Dusky and Creole darters still lived in the gravel shoals that wagons could cross in late summer. It was winter now, and behaving as it should, the bayou left its banks and covered the floodplain 2 miles wide between the red clay hills. Only the leafless crowns of overcup and willow oaks broke the surface of the adjacent flats. Understory water elms and mayhaws slept the season away completely submerged. The captain used the russet-feathered cypresses that lined the banks as channel markers.

One mile upstream the paddle-wheeler passed White's Ferry, closed for the season, and soon after churned over the drowned wreck of the *Rosa B.* In succession, local landmarks were washed in the boat's wake—Catfish Slough, Long Reach, Wolf Brake, Cross Bayou, Bayou Choudrant, Holland's Bluff, Eagle Lake, Old Mills, and finally the destination of the day—a spring-fed creek entering from the east side known as Rocky Branch.

The leader of the expedition, C. B. Moore, was searching for a man who owned land nearby and whose last name was spelled in part like the ancient word "Ouachita." Moore had been informed by his scouts that evidence of aboriginal sites was located on the man's property. The man himself was reportedly an Indian.

Moore was an archaeologist of sorts. He roamed the Southeast plundering Native American mounds and burial grounds searching for artifacts, especially ornate pottery and bones dug by his crew of laborers. Although modern scientists would condemn his destructive techniques, some value would come of his published works.

Moore found his man whose first name was Rufus. He was the patriarch of a local clan subsisting on the fruits of marginal soils and a fickle swamp. His background was clouded by time and suspicion of strangers. Of his ancestors, little was known other than that his father, a private in the 31st Louisiana Infantry, was paroled at the fall of Vicksburg and walked back to this swamp.

Rufus led Moore to areas the family called the big and little Indian camps. Moore described them as humps and rises in a field near the bayou. The laborers dug into them with results that disappointed Moore. They found no intact pots or burials, only broken potsherds, shells, fire-cracked rocks, and small, barbed dart points—worthless in the science of the day.

When questioned, Rufus had no knowledge of the former inhabitants of the Indian camps. He could not have. Later analysis of the site would reveal the occupants to be members of the Coles Creek and Plaquemine cultures, which flourished between 600 and 1,200 years ago.

The drama of the scene on this day was an enigma unappreciated by the players. A Harvard scholar of European origin impatiently scurried about giving orders to men of African descent in the name of science and glory. Ghosts of an American race long vanished drifted among the adjacent trees. Rufus stood barefoot in his fallow cornfield somewhere in the middle of a cultural stew. He could no more imagine the lives of the other actors than he could have a future great-grandson with the same surname who would write about the event for a book called *Bayou-Diversity*.

Autumn Colors

Every autumn a multitude of people in the Northern Hemisphere contributes billions of dollars (that's billions with a capital B) to local economies in order to look at brightly colored leaves. The attraction, when green leaves of hardwood trees turn brilliant shades of yellow, orange, red, and purple, is a result of chemistry pure and simple. Well—simple to a chemist maybe. Actually, the pigments that create the bright colors are already in the leaves in mid-summer long before the pilgrims begin to think of foliage tours. They are masked then by the dominant, green pigment chlorophyll. The trigger to begin the change is called photoperiodism—the length of day and night. As days become shorter and cooler, the production of green chlorophyll declines and the once concealed bright pigments are revealed. The actual autumn colors of the leaves depend on the type and amount of pigments present once the chlorophyll is gone and other chemicals, such as sugars and tannin, in the leaves.

Some species, including hickories, birches, red bud, and tulip poplar, are always yellow, never red. Others, like dogwood, sweet gum, black gum, sumac, oaks, and maples, are usually some shade of red or purple, but may also be yellow.

The intensity of autumn colors is greatly influenced by the weather. Low temperatures and bright sunshine destroy chlorophyll. These same conditions and dry weather enhance the production of the colorful pigments. As a result, the brightest fall colors occur when dry, sunny days are followed by cool, dry nights.

Leaves fall in the autumn as part of a tree's natural lifecycle. It sets the stage for winter dormancy, a time of decreased metabolism in the tree when all physiological processes are geared toward surviving cold temperatures and short days. The bright colors have no apparent biological significance, merely being byproducts of chemical changes leading up to dormancy. Likewise, mass human allure to the beautiful leaves has no perceptible biological explanation.

Contact I

A perspective of the prehistoric human/wetlands interaction in what is now Louisiana leans heavily to the sciences of geology and archaeology. Most evidence leads us to believe that the first humans reached this region by the Late Pleistocene period, 10,000 to 12,000 years ago. It would be difficult for us to imagine their environment. The climate was cooler, sea level was lower, and grasslands were the predominant habitat type. Giant ground sloths, mammoths, horses, and other extinct animals roamed the prairies. Wetlands were a much less dominant part of the landscape and likely concentrated wildlife and their remarkably efficient human predators.

As the Ice Age ended, the climate warmed and Louisiana became humid and heavily forested. Grasslands shrank to relict prairies, and many land surfaces were buried under sediments washed toward the Gulf by the Mississippi River and its tributaries. Gradually, the landscape came to resemble that known by the first Europeans to visit the area in the sixteenth century. Prior to this time, however, the Native American culture continued to evolve, leaving evidence of sophisticated societies dependent on the surrounding wetlands. Hunting continued as a basic pursuit, but emphasis changed to bear, deer, and small game common to bottomland hardwood forests. Waterfowl, fish, turtles, alligators, and freshwater clams, all wetland-dependent species, were important in the diets of these people. There is evidence that native wetland plants, such as marsh elder, lamb's quarters, giant ragweed, pigweed, smartweed, and canary grass, were cultivated prior to the introduction of maize. Archaeological artifacts reveal that spiritual ties to wetland species, such as owls, frogs, eagles, cormorants, shoveler ducks, and roseate spoonbills, existed throughout the prehistoric period. The wetlands-dependent culture at Poverty Point in West Carroll Parish is thought to have had a major influence throughout the Midwest and Southeast.

Wetlands, their regulation, significance, and status in Louisiana, are often debated. Regardless of the future outcome of ongoing debates, there is no denying that wetlands have configured the prehistory of the state.

Contact II

In 1542, Hernando de Soto's plundering expedition passed through Louisiana leaving only hints of the changes his fellow Europeans were to bring. It would be another two hundred years before the vanguard, Frenchmen this time, established a permanent toehold for Europeans in northeast Louisiana. I once lived in Madison Parish and drove to work in Vicksburg. I would often stop on the bluffs overlooking the Mississippi River and imagine the year to be about 1785 and my task to get to Fort Miro in what is now Monroe. Just what would this country be like?

Reports indicate the lands between the Mississippi and Ouachita to be empty of people at the time except for infrequent transients plying the salt trade and occasional Indian hunting and gathering parties. Most aboriginal earthworks, including Indian mounds, had been abandoned for hundreds or even thousands of years. Heading west from the natural levee of the Mississippi, most writers depict a dense forest of bottomland hardwoods with a nearly impassable understory of cane and palmetto. Important mast-producing trees were dominant, with pecan, cherrybark, and water oak on higher sites, grading into nuttall, willow, and overcup oak in lower areas. Giant sweetgums were common. Cypress and buttressed ashes lined the bayous, sloughs, and brakes. After crossing Bayou Macon, one would begin to notice subtle changes. Dense forests gave way to open meadowlands interspersed with clumps of mixed hardwoods and strips of forests along streams. Here, water tupelo replaced the ashes in sandier soils. The waterways were clear except during floods and supported excellent fisheries. Indeed, many people alive today remember the crystalline waters of Tensas River, Bayou Macon, Boeuf River, and Bayou Lafourche.

Wildlife populations were very different. Large predators were still at the top of the food pyramid. At the time, there were more bear, turkeys, ducks, alligators, and songbirds, and less deer, geese, beaver, cormorants, and cowbirds. Wildlife present then but absent now included bison, red wolves, cougars, ivory-billed woodpeckers, Carolina parakeets, and passenger pigeons. Wildlife present now but absent then includes coyotes, starlings, cattle egrets, armadillos, nutria, carp, and fire ants.

Any assessment of this imaginary trip should be conducted with the knowledge that change is a natural part of all ecosystems and occurs with or without humans in the equation. It is the rate of change that differs with the presence of man.

Contact III

Many of the white pioneers who first settled in the hill country of north Louisiana came from Alabama, Mississippi, and Georgia. Some came overland in covered wagons and oxcarts via Natchez or Walnut Hills. Others waited for the seasonal floods and arrived on steamboats after an upstream journey on the Mississippi, Red, Black, and finally Ouachita rivers. They disembarked at Monroe, Loch Lomond, Ouachita City, and Alabama Landing. In many instances the red clay hills, clear streams, and virgin forest resembled the lands from which they migrated. This familiarity enhanced their chances for survival in a wilderness setting. Being almost totally dependent on the natural resources of their environment, they welcomed well-known species of plants and animals. The variety of trees was particularly important.

In the 1830s and 1840s, forests in the hill country were comprised of pine and a great diversity of hardwoods. Since the first European settlements on the eastern seaboard, hardwoods provided essential day-to-day sustenance. Uses were often species specific. White oak provided barrels, shingles, baskets, wagon tongues, and axles. The tannin in southern red oak was used for tanning hides into leather. All oaks were used for fuel and lumber. Red maple yielded furniture and cabinets. Bowls, dishes, wedges, and mauls were made of hornbeam. Dogwood, one of the hardest woods in America, supplied pulleys, mallet heads, yokes, and rake teeth. The tendency of black gum to develop hollows made it useful for water pipes and beehives. The heartwood of black locust lasts up to fifty years in contact with the soil and yielded fence posts and bridge timbers. The fruits of some trees, such as hickories, walnut, and persimmon, contributed seasonal provisions to often-sparse larders. Pharmaceuticals were rendered from toothache tree, sweet bay, willow, and others.

Today the hill country of north Louisiana is still covered with trees, although it should never be confused with the natural forest observed by the pioneers. Monotypic stands of genetically altered, commercial pine trees are the norm. On a landscape scale, the many species of hardwoods are uncommon at best. If the pioneers had encountered today's forests, they may have urged their mules on to Texas.

Front-porch Safari

On a recent sunny afternoon I sat on the front porch of my rural house and daydreamed of a trip to Africa to view the renowned wildlife spectacles. After a while I became aware of a flurry of activity on a scale far from that of elephants at a water hole but no less dramatic. From a rotten log somewhere in the woods in front of my house termites were swarming. Triggered by the rhythm of their biological clocks, they emerge on newly developed wings on a warm day following a spring rain. At first I thought they were ants, but a close examination of one that lit on my rocking chair revealed two pairs of equal-sized wings and the lack of a waist—the identifying characteristics for termites. Thousands of them drifted in from the west in a migration comparable to that of wildebeests on the Serengeti Plains. As the vast herds of African herbivores are always in the presence of predators, so did these consumers of plant materials attract carnivores in search of easy protein to ensure their own survival. Instead of lions and hyenas, attacking squadrons of dragonflies thinned the ranks of the drifting termites. They were easy pickings for predators with names like great blue skimmer and pondhawk. Lacking maneuverability, the termites could only rise from their source log on weak wings and depended on the breeze to carry them along. They lit on every surface they touched. Those hundreds that landed on the porch and nearby sidewalk spawned another drama, this one involving terrestrial predators in the form of the local lizards. The three resident species—anoles, fence lizards, and five-lined skinks—were already in the midst of a tumultuous existence as each individual tried to maintain a territory for breeding purposes on a favored section of the porch or chunk of petrified wood. This involved continuous posturing via push-ups and flashing of gular pouches along with perimeter patrols. Chaos ensued with the sudden presence of the termites. Territories were temporarily abandoned, and trespass was rampant as both inter- and intraspecific competition was focused on gobbling down the most termites. My enjoyment of this front-porch safari was tempered by the sudden thought that hundreds of these cellulose-seeking termites were landing on my wooden house.

Myths and Misconceptions

Myths and misconceptions concerning Louisiana wildlife flourish. Many have been handed down through generations and have origins in puzzling encounters before basic scientific facts had been gathered. Such was the case when an early writer described the habits of denning bears in Louisiana. At the time, it was not understood how bears could survive in a den all winter. In 1801, C. C. Robin wrote, "He feeds himself by licking his paws. Naturalists contend that his paws are provided with glands always imbued with a milky substance. It may be supposed then that this immense quantity of fluid fat which the bear has accumulated during the fruit season runs and increases by means of these glands." Quite creative!

It was once thought that swallows hibernated by burrowing into the mud. This idea may have emerged as a result of their habit of gathering in large flocks over marshes and mudflats to feed on insects, just prior to disappearing across the Gulf of Mexico during migration. Also, the notion that raccoons wash their food before eating derived from their manner of foraging in every available puddle of water.

The origins of some myths are harder to pinpoint. One in particular involves persistent reports of black panther sightings in Louisiana, this in spite of the fact that there has never been a documented case of a black cougar anywhere at any time in history. Black leopards live in Asia, Africa, and zoos, but no black cougars. Don't try telling that to a believer.

The village of Mer Rouge in Morehouse Parish is said to have been named by early explorers who found a prairie there covered with red clover, which resembled a red sea, thus *Mer Rouge*. The reality that red clover is a native of Europe and was not introduced to this area until many years later was overlooked.

During the 1950s and 1960s, white-tailed deer were captured

and transported from Wisconsin to bolster small, remnant herds in Louisiana. Hunters believed them to be larger and darker than our native deer. In fact, the color of a deer's coat is naturally reddish-brown in the summer and bluish-gray in the winter. Still, we hear of a few Wisconsin "blue-bucks" being killed every season, forty years after the introduction of deer identical to those already here.

Scientists are not immune to misconceptions. Until recently most waterfowl biologists thought that ducks migrated south in the fall, basically stayed put on the wintering grounds, and migrated north in the spring. Not so. Ducks carrying radio transmitters have proven that long-distance commuting is common throughout the winter. The pintail that feeds in a rice field east of Monroe on Saturday may have been at coastal Lacassine Refuge on Friday and be in central Arkansas on Sunday before heading back to the Gulf. Two hundred- to 300-mile jaunts after they come south are not uncommon.

Scientific advancements continue to dispel myths, resulting in no need to lose sleep over those snapping turtles lurking on the bottom of your swimming hole. They won't really hold on to your toe until it thunders.

Big Bend/Diversity

As I write this *Bayou-Diversity* story, I am overlooking the muddy roiling Rio Grande nearly 1,000 miles southwest of Louisiana. There is little that can be called civilization for 500 miles to the south. Indeed, much of the looming Sierra del Carmen Mountains in Mexico has never been mapped. To the north, the small village of Marathon, Texas, lies across 100 miles of Chihuahuan desert. Between here and there a road is closed because three mountain lions have recently behaved in a manner bolder than the comfort level of patrolling rangers. Two days ago as we hiked along a remote trail, a limb snapped 15 feet overhead and revealed a black bear casually dining on fruit of a madrone tree. It is autumn; the thermometer only reached 98 degrees Fahrenheit today. There was a scorpion in a cook pot in the bunkhouse the first night and a nest of baby desert mice in the oven—the kind whose picture adorns the hantavirus warning poster on the back of the door.

For several days we have tramped across the desert and mountains with a hand-held space-age gizmo that uses orbiting satellites to help us locate random points. At each site, detailed measurements of vegetation are gathered as part of a larger study to better understand birds of this region. Most plants have thorns, needles, or spines. I have been vaccinated from head to toe. Lechuguilla and cholla are the worst.

Although it's hot, there is no humidity, and I've been much less comfortable in Louisiana. The terrain is moonscape rugged but no more impassable than an overflow swamp. Cacti abound, but poison ivy is rare. Scorpions and diamondback rattlesnakes demand caution, but so do red wasps and cottonmouth moccasins in Louisiana. I've not been bitten by a mosquito here either. It's different, and it's the same. There is no bayou here; the parallel is diversity.

Mud and Muscadines

Having been away from Louisiana bayous and rivers for a long time and then returning once in late summer, I walked down to the water's edge and was suddenly overwhelmed by the appealing odor of mud. It was a memory aroma. Conjured up in the cerebrum via the first cranial nerve when odor molecules wafted across my olfactory neurons, distinct images of boyhood experiences panned across the screen of thoughts—seining a sandbar on a stormy night, checking trotlines at dawn with great anticipation. Only the smell of certain mud evokes these recollections. Hard to describe, it is pungent with a bit of putrid decay and very site specific. That the odor of some mud is alluring, even haunting, defies straight-line cause and effect explanation. I like to think that a gene is involved, perhaps one harbored by my great-grandfather when he floated cypress sawlogs down Bayou D'Arbonne, or yet more distant ancestors who dug peat from Scottish bogs or marked their riparian lifestyles with enduring shell middens along southern waterways.

Another group of biologically secreted odors known as pheromones triggers social and physiological responses in members of the same species, whether honeybees or humans. Depending on the type of pheromone, the resulting behavior can be fight, flight, feast, or mating. Apparently some plant odors are similar to human attractant pheromones, as demonstrated by the floral motifs in most perfumes. The multibillion-dollar global perfume industry produces thousands of products in efforts to satisfy individual tastes. For me it is the fragrance of ripe muscadines, sultry and musky. This wild grape native to the warm, humid South is sleek of skin when mature, round, and firm. Accordingly, to temper passions, wine of muscadine should only be sipped over ice, a vintner's sacrilege.

Odors routinely splash around in our minds directing subconscious behavior and thoughts like a hidden conductor. They warn of danger, help mold first impressions of new acquaintances, new apartments, or old cheeses. Smells drive emotions that crystallize into moods of nostalgia, procreation, and harmony. Such is the mighty power of bayou mud and ripe muscadines.

Blackberries/Civil War

Corporal Rufus Kinsley of the 8th Vermont Regiment wrote from south Louisiana on April 1, 1863: "Co. returned to Bayou Boeuf, had little to eat but blackberries for three days." Of all the plants mentioned in Civil War diaries, journals, and letters, the blackberry was one of the most common. It was so prevalent and so widespread that there is little doubt this wild fruit was an important part of soldiers' diets at times.

Blackberries, members of the rose family, are native to Europe, Asia, North America, and South America and consist of hundreds of species. Dewberries are in this group, and all have been used by humans for food and medicinal purposes for thousands of years. Today many cultivars of wild varieties are in commercial production. Traits such as disease resistance, thornless stems, and large fruit have been purposefully selected. Chemicals within blackberries are now studied for their anticancer properties and ability to scavenge free radicals.

Private Theodore Upson of the 100th Indiana Volunteers near Atlanta on August 13, 1864, wrote: "Our boys are living on fruit diet mostly now. The blackberries are so thick in the abandoned fields that one can pick a ten quart pail full in a few minutes. The boys make puddings, pies and evry thing they can think of."

Union Brigadier General Alpheus Williams wrote his daughters from Warrenton Junction, Virginia, on July 27, 1863: "The fields now are covered with the largest kind of blackberries, both the vine and the bush kind. We have been surfeited with them. For miles and miles in every day's march since crossing the Potomac the fields on both sides of the road have been at every halt, covered with men gathering these berries."

The soldiers were not alone. Along with a host of bird species, bears, foxes, coyotes, mice, and box turtles are attracted to the sweet berries. Deer and rabbits browse the shoots and with others animals find refuge in the thick brambles.

Blackberries

Captain William Seymour of the 1st Louisiana Brigade, near Hedgesville, Virginia, wrote in his diary on July 21, 1863: "Our men were halted in an immense field of black berries, in which the whole Division regaled themselves. The troops declared that it was merely a foraging expedition and that Gen. Early had marched them there to draw rations of blackberries—rations of bread and meat being quite scanty in camp."

The term "ecology" is often defined as the interrelationship of an organism with its environment. Although not a typical example, humans in blackberry patches during the Civil War were nonetheless an ecological association.

Autumn Flight

It's hard for me to believe that at the time I was his student, he was younger than I am now. I never thought of him as young, certainly not as young as I am today, but I never considered him old either, even when he died at the age of 65.

He seemed to follow me from bayou to bayou, flitting in and out of my life for more than twenty years much like his passion and the focus of his vocation. Dr. Tom Kee was a professor of biology, and while he enthusiastically taught the life histories of mammals and parasites, he lived birds. It was impossible to be in his presence for sixty seconds and not recognize this as fact. To be one of the hundreds of survivors of his spring ornithology class was to know it as a universal truth.

Cloak him in feathers and you would have a white-crowned sparrow in spectacles. Slight of frame with a shining pate, he proclaimed the joys of Class Aves from every available perch. Like this sparrow he was conceived in the North Country and migrated south. Born in the Upper Peninsula of Michigan, he schooled in that state and Arkansas and finally established a permanent territory around Northeast Louisiana University at Monroe. Unlike the sparrow's, his trip was one way in spite of often professed yearnings to spend the autumn of his life on natal grounds.

"Birds according to Kee" was not a subject to be taught in a laboratory with stuffed, silent, still specimens. He was of the field naturalist bent and rallied groggy students in the predawn darkness for forays into the magic of spring migration. Here was the essence of life as expressed in color, sound, and movement. Here were the mysteries of adaptability, flight, and navigation. These lessons were taught simply and with such natural fervor that few students could resist imprinting on his enthusiasm. I reveled in riding his slipstream.

After basic ornithology, I completed his courses in field ornithology, upland gamebirds, and waterbirds and marsh birds. I fulfilled requirements for a problems course to survey wintering bald eagles under his supervision. Gleefully, he led us on treasure hunts to Cajun marshes, Texas beaches, and the Rio Grande Valley. Wealth was measured in surf scoters, chachalacas, and elf owls.

In the summer of 1971, along with other undergraduates I followed him to a secluded field camp on the White River in northern Arkansas to spend six weeks studying the flora and fauna of the local environment. It was a fledgling naturalist's paradise, and along with bats, 2-foot-long salamanders, and jewelweed, I discovered Amy. As we performed courtship rituals that led to a steadfast pair-bonding, Dr. Kee's curfews and patience were stretched. He responded with his omnipresent, powder-dry sense of humor.

Amy and I departed for graduate school unconsciously laden with his lessons. Later, after I began working for the U.S. Fish and Wildlife Service, he continued to drop in on our life from time to time and place to place. In the marshes of Lacassine National Wildlife Refuge it took him only ten minutes to find the vermilion flycatcher I had sought for weeks. At Tensas River National Wildlife Refuge he surveyed millions of blackbirds during the annual Christmas Bird Count. He monitored endangered red-cockaded woodpeckers for us at D'Arbonne National Wildlife Refuge.

He retired in 1991 and decided to remain in place close to his grown children. As a retirement gift one of my sisters, a colleague of Dr. Kee at the university, organized a fund drive for an all-expense-paid trip to the birding mecca of Costa Rica. Donations from former students and other friends poured in. My brother, Dr. Kee's last graduate student, accompanied him south and relayed his mentor's youthful delight on discovering resplendent quetzals, black guans, and three-wattled bellbirds. Soon after, his health began to resemble the undulating flight of a flicker, yet his lust for birding endured. In the fall of 1994, he was invited back to the university to teach part-time, but his heart failed two weeks later.

On September 24, 1994, five of us, comrades, students, co-workers, gathered at the Monroe Regional Airport to effect his last wishes. The low-wing Piper carried us southwest 200 miles to a small live oak chenier hugging the vastness of the Gulf of Mexico. It was his favorite spot in the world. In the spring, thousands of weary, weather-beaten migrants saw it as first landfall after a nonstop flight across the Gulf. Their dramatic fallouts into the wind-sheared vegetation offered some of the best birding in the world. Dr. Kee hailed their return here year after year.

All five of us were busy in the last moments. There was a quiet prayer. The pilot concentrated on the aircraft as we slowed and settled to within 400 feet of the treetops. The navigator counted off the seconds and monitored the ominously approaching radio towers. The back door was opened, and the device was lowered beneath the wing into the turbulence. At precisely 10:30 a.m., ripcords were pulled and a fine plume of ash settled into the ancient oaks. A flock of roseate spoonbills flared in the sunlight as we banked to the north.

There's a good chance that I will encounter him again. Several years may pass as nutrients flow into the soil, through Louisiana irises and gaudy caterpillars, and eventually into vibrant flocks of neotropical migrants. One thing is certain—I'll be birdwatching.

Bayou Steamboats I

Anxiously they lay in wait—*Fox, Sun Flower, Silver Moon, Persia, Sydonia, Red Chief, Lucy Robinson, Young America, W. W. Farmer,*

Kate Dale—their captains eyeing the slowly rising waters of the Ouachita. During the last half of the nineteenth century, these steamboats and others would gather at the mouth of Bayou Bartholomew just above Sterlington usually in late November or early December. When winter rains raised water levels high enough for navigation, the normally quiet bayou seethed with activity as the steamboats churned its waters in a dash to serve the wealthy cotton planters and small communities along its shores. In varying degrees the activity continued until May or June, when the bayou again receded to a winding, crystalline stream often no more than knee deep.

The boat captains, deck hands, and passengers never realized that as they traveled Bayou Bartholomew they passed over the most biologically diverse area in terms of fish species in the state of Louisiana. Even today few people know that researchers have found more kinds of fish in Bayou Bartholomew than any other body of water in the state. At least 101 species are known to inhabit the bayou. They range from common gamefish, such as bass and bluegills, to less celebrated but intriguing crystal and stargazing darters, pugnose minnows, and redfin shiners.

When the aquatic biodiversity of Bayou Bartholomew is compared with that of other area streams, such as Bayou Lafourche, Bayou Macon, and the Tensas River, extreme differences are noted. Bayou Bartholomew has twice the number of species as the others. The reason lies partially in the fact that, unlike the others, Bayou Bartholomew has never been significantly channelized or dredged. Habitats such as gravel beds and riffle areas necessary in the lifecycle of many fish remain intact. Water quality is relatively good. Diversity in the other bayous has gone the way of the steamboats.

Bayou Steamboats II

Had you been sitting on the bank of Bayou Bartholomew several miles below Bastrop on the afternoon of December 13, 1857, you would have heard her piercing scream long before she came into view. Heavily laden with cotton bales, the steamboat *W. W. Farmer* eased cautiously along with the current. Water levels were rising but still low. The first boat of the season had been able to enter the mouth of the bayou from the Ouachita River only nine days before. The *Silver Moon, Lucy Robinson,* and *Young America* were also trading up and down the sinuous stream so narrow that two boats could barely pass. Navigation was treacherous. In two months the *Red Chief* would lose a smokestack here on an overhanging limb. In two years the *Princess* would lose a boiler and liberate the souls of nearly two hundred passengers. Lookouts on the *Farmer* watched for sparks from the stacks and quickly doused any that settled on the incendiary cotton bales.

As the *Farmer* came upstream a couple of days earlier, her steam whistle had announced stops at the numerous plantations and other landings. Mail, freight, and Christmas orders, mostly via New Orleans, were off-loaded. Wealthier customers received crates of iced oysters fresher than those in today's local markets and fine cloths from Europe. Apples and oranges were also treasures in this emerging pioneer world. Yeoman farmers picked up staples, such as flour and coffee, and necessities like plow harnesses and tobacco.

This scene was being repeated in smaller streams across northeast Louisiana. Steamboats worked Boeuf River to Point Jefferson, the Tensas River as far as Waverly, and Bayou D'Arbonne above Farmerville. Flood control and navigation projects on these waters were decades in the future. Commerce and life in general revolved around the cyclic, nourishing floodwaters.

The concept of manifest destiny as it applies to nearsighted efforts to control natural processes eventually overwhelmed most of the smaller waterways. Dredging and straightening converted much of the Boeuf and Tensas rivers into manmade plumbing systems. Bayou D'Arbonne was dammed. Some good came of these projects but rarely at the levels predicted. Prophesied benefits to commerce seldom materialized, and hundred-year floods ignored the calendar with increasing frequency. Aquatic wildlife populations took a beating. Several taxa of fish disappeared when critical streamflows, water quality, and bottom substrates were altered.

There are still semblances of wild bayous in the area. The captains of the *W. W. Farmer* and other such vessels would recognize Bayou D'Arbonne below the dam, Tensas River within the national wildlife refuge, and the first few miles of Bayou Bartholomew. There is no good reason not to keep them so.

Names

Humans have been putting names on plants and animals ever since the first caveman found it advantageous to convey to his mate the difference between a cave cricket and a cave bear. As our species developed culturally and interactions between groups who spoke different languages increased, the problem surfaced again. The folks on the other side of the mountain were peddling rugs from an animal they called moose, but which were known to the prospective buyers as camels. It was a problem of conflicting common names, and it got worse when scientists came along. A Swedish fellow named Carl Linnaeus began working on the issue in the 1740s by developing a system to give a two-part Latin name to every plant and animal on the planet. By doing so, when scientists in Japan, Somalia, and Brazil heard someone from Rocky Branch, Louisiana, speaking of *Didelphis marsupialis,* they knew he was talking about the North American opossum.

The system has been fine tuned over the years and works pretty well for scientists with only occasional squabbles between the splitters and lumpers—those who like to divide species based on minute differences, and those who throw everything similar under one Latin moniker. Still, the layman who did not wish to memorize thousands of tongue-twisting dead-language words and match them with plants and animals in his personal memory bank was in the dark, or at least thought to be so by those who claimed to be in the light. The bright guys then made a list of standardized common names in an effort to force abandonment of colloquialisms. Many have refused to go there, though, and examples abound just in the bird world. The term "di-dipper" survives in the face of the official pied-billed grebe. Water turkeys still swim with double-crested cormorants, squealers roost with wood ducks, and buzzards yet soar with almighty vultures.

Many years ago as a budding scientist I might have corrected the misguided individual who called a wood stork a gourdhead, but not today. It feels better to sit on the bayou bank in the shade of a bitter pecan (not a water hickory as the books say, or even worse *Carya aquatica*) while fishing for chinkapins (not redear sunfish or *Lepomis microlophus*), and listening to the songs of wild canaries (instead of prothonotary warblers or *Protonotaria citrea*)—sort of like a caveman, I suppose.

Pasaw Island

The religion of ancient Celtic peoples included a concept known as "thin places." In simple terms, a "thin place" is a physical place on the landscape where the veil between this world and the next is thin. It is a place where one easily senses the spiritual existence of Heaven, and signs of its being are almost palpable. I am convinced that "thin places" exist today for those who are receptive to their mysteries. Examples could include the sanctuary of an old country church, a scenic overlook in the mountains, a deserted beach, or even a quiet nook in a city park. Such places vary with the perceptions of the beholder.

I know of one "thin place" deep in the heart of a Louisiana swamp. Local people refer to it as Pasaw Island, allegedly because someone's "pa" once "saw" a bear there. It is a true island only during the primordial rhythms of backwater floods. At other times it is prominent in its height above an otherwise flat, heavily forested swamp. Rising 30 feet above overcup oak flats in an area where a 2-foot difference in elevation drastically changes hydrology and vegetation type, the island is a physiographic aberration. It exists because the bayou that created the swamp by eroding away a Pleistocene terrace missed Pasaw Island in its meanderings across the floodplain. A quarter-mile long and half as wide, it resembles in shape the overturned keel of Noah's ark.

Trees on Pasaw Island are not of the swamp. Many are senescent and dying slowly as a result. Loblolly pines, black hickories, and white oaks crown the top. A grove of huge tree huckleberries with twisted, naked limbs straight from a fairy tale forest clings to the east slope. The flowers of a single giant dogwood reflect ethereal light on early spring mornings. Even the soils are different from the dark, heavy clays of the swamp. Rootings of Johnny-come-lately armadillos reveal red, sandy-clays with iron concretions. Evidence of humans contemporary with the Celtic tribes lies scattered in grog-tempered potshards. It is usually quiet there, provoking contemplation . . . and even at only 30 feet higher than the swamp, it is light-years closer to Heaven.

Hunting

Along the Louisiana bayous, hunting is a traditional recreational pursuit for thousands of people. The economic impact of this activity is valued in the millions of dollars. A multifaceted industry has developed to supply hunters with guns, ammo, clothing, transportation, calls, blinds, lures, decoys, dogs, and a host of other equipment. Getting geared up for a local hunt is as easy as stopping by the nearest sporting goods store or placing an online catalog order. It wasn't always the case, though. Consider the following contract, which may have been for one of the first outfitted hunts in the Ouachita Valley. The setting was colonial Louisiana. Even though Spain had recently acquired the region in a secret treaty, the French were still de facto rulers in 1764. The contract reads:

> Before the register of the Post of the Natchitoches personally appeared Andre' Rambin, inhabitant of this post, and Joseph Gallien, hunter, who have agreed as follows: that the said Joseph Galien contracts to go

to the Black River above the Houachitas, with a Negro man and an Indian woman to hunt buffalo, bear, and deer until the month of February of next year, seventeen-sixty-five, from which hunt the said Gallien promises to deliver half, be it in skins of deer, bear oil, liver and tongue of buffalo or salted meat, to the said Rambin, in consideration of which he will furnish Gallien twenty-five pounds of powder, fifty pounds of balls, a flintlock musket or a better kind if such can be found, a hogshead of salt, a copper cauldron of about nine and a half pounds, twelve butcher knives, fifty gun flints, twelve ramrods, two axes, an adz[e], a mosquito bers [bar] of twelve ells of brin, fifteen pots of tafia, fifty pounds of wheat flour, twelve pounds of sugar, twelve pounds of husked rice, three barrels of kernels of corn, eight pots of bear oil, three hundred livres royal silver for the hiring of the Negro which Gallien will deduct from Rambin's share at New Orleans where Gallien is to take the boats loaded as aforesaid with deer skins, bear oil, salted meat, and buffalo tongue, after which each will draw out his half of the supplies left from the hunt, in witness whereof the said Rambin has signed and the said Joseph Gallien has made his usual mark in the presence of Monsieur le Chevalier de Macarty, commandant, and Monsieur de Santelette, merchant, done at Natchitoches the 16th March 1764.

As an anecdotal follow-up, it should be noted that only the quality of hunting gear has changed. The quantity for a normal outing remains the same.

Halloween Visit

People visit Louisiana bayous and forests for various reasons. William came up out of the D'Arbonne Swamp on Halloween day and badly frightened the secretaries in our remote office. I received their distress call saying there was a man at the office causing a disturbance. When I arrived I found William—medium height, stocky and scruffy, about my age. He had a peculiar stare; his pupils were tiny dots, and he smelled of alcohol. He did not talk much at first but handed me his car keys, driver's license, and a newspaper clipping. The clipping was a month-old obituary notice. William said it was about his favorite brother. William was reluctantly cooperative when I moved him outside and searched him for weapons. Over the next hour, a drama played out that seemed at the time surreal. As I tried to tease out the circumstances leading up to William's unsettling visit, he was lucid one moment and babbling incoherently the next. I learned that we had attended the same high school, and we talked of teachers whom we remembered. Occasionally, he would double over in apparent pain and clutch his shoulder. He said a boating accident caused the injury, which he could not afford to have treated. William said he didn't want to be alone now and that no one cared about him. His behavior became more erratic. At one point he asked to retrieve medication from a lunch box in his car. I should have known better. I did know better. He came out with an eight-inch long hunting knife—and presented it to me as a gift because I had said I cared about him—this after declaring that he had no weapons of any sort in the car. He rambled about, and I had a hard time containing him. I almost arrested him on a couple of occasions and had just cause to do so. Finally, I had two chairs brought outside and an ambulance and deputies summoned. William and I sat in the chairs on the sidewalk facing each other and held hands as members of the public walked around us going in and out of the government office. Their sideways glances were not discreet. William smoked and described the intricacies of Sig, Glock, and Browning pistols. He said his family had been trying to have him committed to a mental institution—"nut house" he called it. Before the ambulance came and carried him away, William said that he had been living in the forest for several days. He professed his love for the swamp, its solitude, unknowable mysteries, and therapeutic values. On past occasions, nature's healing virtues had helped him through the valleys. This time the restorative was elusive, the shadows too long.

Unnatural Selection

A while back I wrote about the dispersal of white oak acorns and how they fell onto my new steep, metal roof and launched themselves over the edge into my yard. I did not realize at the time that I would soon involuntarily mimic the acorns by launching myself from the peak of this roof, down the 7/12 pitch, and onto the sidewalk below. The laws of physics functioned according to established equations, particularly acceleration—both positive and negative—and gravity. The density of mass was tested, and, as might be expected, that of concrete exceeds that of older human bone. Density = Mass/Volume, you remember. This incident stimulated me to the extreme and continues to stimulate the local healthcare-related economy. The fall initiated a chain of events that involved, among others, an exasperated wife, a perplexed state trooper, and a very talented orthopedic surgeon. I don't really have time here to go into the details, except to say that because of all the metal screws and other foreign hardware now in my foot, airport security checks will never be the same for me again.

So how does all of this relate to the normal theme of this book, which pertains to natural history? Here's the nexus: a wild duck is to blame for my broken foot. It's true. Two species of waterfowl common in Louisiana build nests and lay their eggs in tree cavities. The hooded merganser is a small, black and white fish-eating duck that produces eggs, which are almost perfectly spherical. This species is innocent of the matter. The other local and very guilty cavity nester is the wood duck. Wood ducks will also readily nest in manmade boxes intended for that use and other manmade devices not so intended—such as my fireplace chimney. This aggravating habit predicates the need to cover my chimney each spring in order to prevent a chimney full of trapped ducks—which has happened, by the way. Thus, no wood ducks—no need to cover my chimney and fall off the house in the process. Or is this really the case? If I had not built my house with its inviting chimney in good wood duck habitat, there would have been no incident. By eliminating mentally unfit wildlife biologists from the population at large, perhaps this was nothing more than a prime example of natural selection at work.

Providence

Less than three months after Union Parish was carved from Ouachita as a new political entity, William McKay died there intestate, leaving behind a grieving widow and 2-year old daughter. In 1839, Union Parish was essentially wilderness and sparsely populated, the surge of immigration by settlers from eastern states just over the horizon. McKay owned a store on the Ouachita River, either at what would later be called Alabama Landing or farther south at Ouachita City, or maybe even at the mouth of Bayou de l'Outre. In these roadless times goods moved efficiently only by water. To settle his estate McKay's widow petitioned the court, and the judge ordered an inventory of the deceased's possessions including the contents of his store. The results are enlightening.

Many of the items would be expected in a wilderness outpost far from civilization: rifles and shotguns with powder and lead for subsistence hunting; fishing line and hooks for the backwater bounties; palmetto hats to turn the sun; molds to produce hog and bear fat candles; broad axes, grubbing hoes; bone and mussel-shell buttons; sugar and molasses; castor oil and patent medicines; playing cards, fiddle strings, and Muscat wine. On the other hand, McKay's customers could purchase imported Irish linen, embroidered lace silk gloves, velvet bonnets, Morocco shoes, fancy soup terrines, decorative chamber pots, Castile soap and tortoise shell combs, bathing tubs, newspapers, reading, spelling, and song books, along with spectacles if needed. No doubt these items came upstream on steamboats via the international port of New Orleans.

Whatever the taste or means of Mr. McKay's customers, there was one article in his shop essential to the survival of anyone of European descent in this raw country. His shelves had an ample supply of mosquito netting—by the yard or as finished bed drapes. Although his customers did not comprehend the undiscovered link between devastating outbreaks of malaria and yellow fever and the hordes of native mosquitoes, they unknowingly contributed to the successful settlement of Union Parish by purchasing McKay's mosquito nets just to get a good night's sleep.

Unwritten Law

Here on this property where we live and that we call Heartwood, there is an unwritten game law. It is "thou shall not hunt within a quarter-mile of the house." The doe that browses just outside my home office window at noon has tempted me on occasion to propose an amendment to this family statute. She and several of her kin are well aware that they are relatively safe here in this small sanctuary, but even like free people around the world they tend to push and test the boundaries of liberty. My binoculars pull in her physical details, as she is only 50 feet away. Body size and a long head with a roman nose indicate that she is an older doe, probably more than 2 years old. Except for a gray muzzle, she has the typical summer coat of reddish-brown hair. A full udder reveals that she is lactating and most likely has a fawn or perhaps twins hidden nearby. They should be about 2 months old now. She seems to be in good shape for this time of year but is not fat by any means. Late summer is the most physiologically stressful season for deer in bayou country. Food plants have become rank and lost much of their nutritional value. Critical hard mast foods like acorns are not yet available. Internal and external parasite loads are highest now. Does are further burdened with the strain of supporting fawns, and bucks are growing nutritionally demanding antlers. On this day I can make out that the doe is purposefully selecting poison ivy to browse—and technically she *is* browsing. Cows graze. Poison ivy is a favored food, and deer apparently are immune to it. Not so for wildlife biologists who naively collect deer stomach samples during routine herd health checks. Nowhere on earth is the poison ivy toxin, urushiol, more concentrated than in a deer's stomach. That I can personally vouch for. My complaint with the resident white-tails is that they don't restrict their diets to the common native plants in my yard. They insist on munching the flower buds from wild azaleas native to the

Appalachians that I have otherwise nurtured with tender care, and my vegetable garden resembles a minimum-security prison. On occasion I have declared to them in a loud voice that the law can be changed, but they seem to know who holds power on the Heartwood Supreme Court.

Rocking Chair

Not long ago a cousin passed along to me a chair that once belonged to my great-great-grandmother. She is said to have brought the chair with her when she came to Union Parish from the bayous of the Atchafalaya Swamp. She died in 1925 at the age of 77. The chair is laden with hints and mysteries of lives past. Just a bit larger than a child's chair now, it was originally a rocking chair but was converted to a simple ladder-back when the rockers wore out. Someone with hardscrabble talent built the handmade chair out of native white oak. Flaking, brown paint reveals bare wood with long, horizontal rays, a characteristic that distinguishes white oak from red. Although simple in style, the chair is not without a touch of refined craftsmanship in the three thin, curved slats that undoubtedly supported the weary backs of subsistence farmers. The seat was originally made from the stretched hide of either a deer or cow. About the time of the Great Depression the seat wore out, timing that may not be coincidental. By then the chair had been passed down to my cousin's mother, and her husband promptly replaced the seat with a rough-sawn cypress board. Saw marks on the board indicate that the lumber was cut with a circular saw blade about 3 feet in diameter. It was not run through a planer, thus forgoing an unnecessary expense in a time when cash was scarce. Someone, probably a mischievous boy cousin with a jackknife, whittled small notches in the top slat when no one was looking. The most intriguing parts of the chair are the front stretchers. These rungs have been worn to nearly half their original diameter by propped feet, and likely by the same person because most wear is on the same part of the same stretcher—a person who favored his or her left foot. Knowing how hard it was to survive on a small farm in the red clay hills, I suspect the mark to be an artifact of worry. On the other hand, Great-great-grandmother was known to revel in the music of her small, round accordion. I prefer to think she rocked the rockers off her chair and then wore the rung through while marking time to a Cajun reel.

Close Encounter

On a sunny July day almost thirty-five years ago I observed a large group of people have an encounter with nature that none of us would ever forget. At the time I was taking a summer break from graduate studies in wildlife science by working on the Trans-Alaskan Pipeline, America's first large, environmentally hypersensitive energy project. Before the pipeline could be built, a haul road was constructed along the right-of-way for equipment, pipe, and worker access. Temporary camps were scattered along the haul road to house the workers. The encounter occurred at Old Man Camp, about 30 miles north of the Yukon River. Like all of the camps, Old Man consisted of connected prefab units and on this particular site was arranged around the edges of a bowl-shaped depression, almost like an angular donut. On this day most of the workers had stayed at camp to hear labor union leaders who were flying in to discuss issues of concern. A podium was set up in the depression, and more than three hundred people crowded into the natural amphitheater. Except for one small opening, they were surrounded on all sides by the walls of the camp. I managed to find a good seat high atop a storage van and waited without enthusiasm for the anticipated boring speeches to come.

As it turned out I was not bored for long, and I don't really remember what became of the speakers. On several occasions in the preceding weeks, a male, sub-adult grizzly, no doubt drawn by odors from the kitchen and readily available garbage, had visited the camp in the short summer nights, leaving behind a trail of mischief. With his long, formidable claws he would pry windshields from trucks, leaving the glass mysteriously intact, to get at a lunch bucket inside. In search of stowed food, he routinely tore the doors from parked buses that carried workers to the jobsite. Having thus earned a deserved reputation as a nuisance and potentially dangerous bear, he was once anesthetized with a dart gun, dangled from a net under a helicopter, and released 100 miles away over the nearest mountain range. He was back in three days. I saw him on a couple of occasions, and unlike the plodding behavior of the abundant black bears his movements seemed charged with nervous energy.

On the day of the union meetings he appeared again. To me the incident seemed to play out in slow motion. The grizzly suddenly came loping through the small opening into the depression. The crowd was packed shoulder to shoulder with nowhere to go. I will always remember the sound. No one screamed, but a long, collective moan swept across the mass of humanity. It was a cry of communal, primal terror. The bear lunged 100 feet into the crowd before he stopped. Somehow a seam had opened and then closed in his wake. From my vantage point I could see the bear was as afraid as the people. He whirled around, parted the herd again, and was gone— all of this in seconds.

This encounter with nature was obviously dramatic for everyone present and left lifelong impressions. Although it is an extreme example, it reminds me that many people in America have almost no experience with nature. City and suburban dwellers alike often grow up and live their lives thinking St. Augustine lawns, pigeons, and red tip photinia are natural components of their ecosystems. Significantly, many topple into a culture that believes humans and nature are somehow separate, that think all environmental problems can be solved with technology—an unfortunate, self-limiting misconception. Close personal encounters with the natural world are the antidotes to ecological naiveté. They don't require grizzly bears. In Louisiana, wildflower gardens, fishing trips, bird feeders, and attentive walks along the bayou will do just fine with time.

5
COLLISION
A Crash or Conflict

Wood smoke from windrows fills the air for miles. The persistent wails of the clearing tractors can be heard plainly from the shop site this morning as Section 25 succumbs. A couple of days ago I saw 3 turkeys standing in a small tract of woods being devoured daily from all sides. No way out but up—the eternal high. The dragons [mills] in Tallulah, Greenville and Natchez belch and demand more. The cats get fatter as my children become destitute.

> —KO Field Diary, 20 July 1984—Tensas River NWR

Waterfowl Law Enforcement work; caught 3 subjects with 47 ducks near [Bayou Lafourche]

> —KO Field Diary, 19 December 1987

Impacts

On the surface it doesn't seem possible. How can we catch all the fish in the seas? Analogies do exist.

Bison, or buffalo, were once the most numerous single species of large wild mammal on the planet. They blanketed the Great Plains of North America and were the life-blood of Plains Indian societies for thousands of years. During the nineteenth century, commercial hunters spurred on by government policies aimed at subduing Native Americans by eliminating their food supplies killed more than 50 million bison. The once vast herds were reduced to a few hundred individuals. In Louisiana bison were common and frequently mentioned by colonial-era writers. The last known individual in the state was killed near present-day Monroe in 1803.

Similarly, passenger pigeons were once the most common bird in North America. An estimated 5 billion passenger pigeons were found on the continent when Europeans arrived. During migration, flocks 1 mile wide and 300 miles long were recorded, which took several days to pass a fixed location. Again, during the nineteenth century, the species went from being one of the most abundant birds in the world to extinction as a result of market hunting and human-induced habitat loss. The last record of passenger pigeons in Louisiana occurred near Mer Rouge in 1903, exactly a hundred years after the last bison.

No doubt people who witnessed the immense herds of bison and flocks of passenger pigeons thought their numbers inexhaustible. Now consider the fish in the seas. Recent studies have shown that within all of the world's large marine ecosystems 29 percent of the species "had been fished so heavily or were so affected by pollution or habitat loss that they were down to 10 percent of previous levels." This meets the definition of a collapsed fishery. Cod have been reduced to between 1 and 3 percent of their natural abundance, and Atlantic bluefin tuna populations have declined 90 percent since 1970. On this course, a global collapse of most commercial species is predicted by 2048 if conservation actions are not taken. Humans do indeed have the capability and at times the indifference to completely eliminate a species from our planet.

Ouachita/Amazon

On a recent trip to the upper Amazon Basin I was able to see the Ouachita River as it appeared two hundred years ago. The time reference could also be labeled as 100 BC—BC being before the Corps of Engineers and their snagging, dredging, and lock-building efforts to domesticate a feral river in the good name of economic development. Riding down the Madre de Dios River in a 30-foot canoe, like a giant, hollow pencil sharpened at both ends, one only has to squint a bit to blur the unfamiliar riparian vegetation of a rain forest into a generic green mass that resembles the bottomland hardwoods along the Ouachita.

The Amazon tributary, unencumbered by levees and dams, is homeless and wanders the jungle floodplain in the manner of what geologists call a braided stream. Today's main channel may be an oxbow lake next week as the river seeks a path of least resistance to carry its combined burden of glacial till, suspended clays, and organic nutrients ever seaward. Point bars, cut banks, and meander loops come and go with seasonal frequency. On the Ouachita such instability is viewed as counterproductive, dangerous, and a challenge to modern riverine engineering.

Human activities along the Madre de Dios today and the Ouachita two centuries ago are similar. In the jungle, scattered small villages of indigenous peoples cling to the high outside banks, their palm-thatched huts not unlike those palmetto-roofed houses of Choctaw along the Ouachita. Small gardens and bounty of the forest sustained them both. Evidence of small-scale illegal logging by poor natives in the form of riverside piles of rough-sawn timbers mar the image of a pristine rain forest. Such activities can be compared to similar practices by Native Americans farther north on lands dubiously claimed by the likes of the Baron de Bastrop and Marquis de Maison Rouge.

Biodiversity along the Ouachita never approached that of the Madre de Dios even in the best of times, a function of the Amazon's isolation from the glacial-driven extinctions of North America. The faunal assemblage there is intact, and while wildlife is still abundant in our region, several important historical players are gone forever. Carolina parakeets don't visit clay licks along the Ouachita like their counterpart parrots and macaws in the Amazon. Great flocks of passenger pigeons no longer break the limbs of oaks in their acorn-feeding frenzies, and Ouachita River fords felt the last sharp hooves of bison more than two centuries ago.

Along many rivers in the Amazon Basin, all of the parts are still present, connected, and humming along according to natural rhythms and processes. If one believes in the repetition of history, this short essay, in a flip-flopped sort of way, may yet be relevant there in two hundred years.

Chestnut and Chinquapin

The greatest disruption to the ecological integrity of eastern forests has been the loss of American chestnut trees. This majestic hardwood species, which towered to 120 feet tall, comprised as much as 50 percent of upland forests from Maine to Alabama. In 1904, a parasitic fungus called chestnut blight was unwittingly introduced into the United States on imported Chinese chestnuts and resulted in the decimation of the nonresistant American species. Chestnut was the keystone species in eastern forests, producing immense volumes of mast in the form of edible nuts that supported a broad diversity of native birds and mammals.

Chestnut was not a native plant in Louisiana. Chinquapin, a smaller, closely related species, does grow in Louisiana uplands. Usually less than 30 feet in height, they resemble chestnuts and produce similar but smaller, edible nuts. Chinquapins also suffer from chestnut blight but still manage to survive in many areas of the eastern and southern United States.

Utilitarian values of chestnuts and chinquapins were similar. The wood of both species is rot resistant and was desirable for posts and railroad crossties. The larger chestnuts were a prime source of building lumber and also used in the manufacture of boxes, musical instruments, tool handles, barrel staves, and caskets. The bark of both species was a valuable source of tannin for the leather industry. Medicinally, a tea made from the roots was recommended as a substitute for quinine. Farmers depended on the fruits to fatten their herds of free-roaming hogs. Humans, too, relished the nuts of both species, parched, boiled, or raw. Sixteen-year old Sarah Wadley wrote in her diary from Trenton, Louisiana, on September 27, 1861: "Yesterday we all went chinquapen hunting . . . We went in the great lumber wagon with four mules . . . first to Mr. Nash's place, where we found 'ever so many' chinquapins . . . we did not gather more than two hours and we had over a bushel."

In Louisiana, chinquapins have managed to hang on in spite of a foreign disease. However, they have not been able to adapt to a second human-induced assault. The conversion of thousands of acres of once botanically diverse uplands to pine plantations is fast sending Louisiana chinquapins down the same one-way road as American chestnuts.

Warming Up I

As one gazes across the allegorical landscape of our country there is still to be seen in all of the many habitats a goodly number of ostriches exhibiting the classic head-in-sand pose when it comes to the acceptance of climate change as a reality. They have yet to yield to the preponderance of evidence, perhaps because of a mindset that defers only to the bears just outside the door. How could a melting iceberg in Antarctica mean anything other than "just one of those things"?

A recent study of plant hardiness zones by the National Arbor Day Foundation brings the boogers into our backyards. The U.S. Department of Agriculture has divided the United States into eleven geographic zones based on the average annual minimum temperature. Each zone is separated from its neighbor by 10 degrees Fahrenheit. Maps have been developed that show in detail the lowest temperatures that can be expected each year in any area of the country. The idea behind creating the zones was to provide information to gardeners that could be used to determine the kind of plants that would grow in their region. Traditionally, Louisiana, with the exception of a narrow fringe along the coast, has fallen into Zone 8, which has an average minimum temperature between 10 and 20 degrees. The coastal area was in Zone 9, with minimum temperatures between 20 and 30 degrees each winter.

In 2006, the National Arbor Day Foundation updated the old maps using the most recent fifteen years of data from five thousand weather stations in the United States. Their results should bring the most recumbent ostriches to attention. The northern boundary of Zone 8 in Louisiana once fell near the Arkansas-Louisiana line. Much of it now lays two-thirds of the way up the state of Arkansas. Zone 9, once restricted to the coastal fringe, now projects north of the city of Alexandria. The trend continues across the country. Most of Illinois, Indiana, and Ohio, for example, have shifted from Zone 5 to a warmer Zone 6. Some areas have warmed two full zones.

In Louisiana, this phenomenon correlates with unusual observations. For many years, severe infestations of the highly invasive Chinese tallow tree were limited to the southern half of the state. Recently, it has invaded every parish. Large numbers of waterfowl that once wintered in Louisiana now loaf the season away in new ice-free areas of the Midwest. Denying the reality of climate change, even in the Bayou State, places the ostriches in ever more precarious positions.

Warming Up II

Climate change as a social hot-button issue heats up and cools with the seasons. Most everyone agrees that the Earth's temperature fluctuates naturally over time. How else do you explain the Ice Ages and fossil palm trees in the Arctic? The rub has been whether or not humans are actively causing the most recent changes. Overwhelming evidence from reputable scientists around the world now make the position of naysayers untenable in their efforts to release humans from culpability.

Over the past fifty years the average global temperature has increased at the fastest rate in recorded history. The ten hottest years have occurred since 1990 and experts think the trend is accelerating. The culprit is carbon dioxide and other pollutants that collect in the atmosphere and trap the sun's heat. Coal-burning power plants and automobiles are the biggest sources of carbon dioxide pollution in the United States. Although Americans make up just 4 percent of the world's population, we produce 25 percent of the pollution associated with burning fossil fuels.

The effects of this artificial and relatively rapid warming of our planet are unknown, but the forecasts don't bode well. The perennial polar ice cap is vanishing at the rate of 9 percent per decade. Resulting higher sea levels and the present lifestyle in south Louisiana are not compatible. Warmer oceans tend to stoke the furnaces of tropical storms that can lead to more hurricanes. The impacts scale down from systems to individual species. Without polar ice, polar bears are in trouble. There is growing concern that because of warming southern reservoirs, those that support trout fisheries in their tailwaters will soon lose this resource. Already waterfowl populations are wintering farther north in direct correlation with warmer winter temperatures. Fire ants are on the move too, pushing ever northward in response to the changes.

The solution is simple: reduce pollution from vehicles and power plants. Technologies exist to build cleaner cars and power generators. Only the willpower of Americans is required.

Warming Up III

Study after study continues to document the phenomenon of accelerated global warming. This research also brings to light new understanding of just how complex the Earth's climate systems are and how unpredictable the consequences can be. What is known for sure is that global warming will not, contrary to intuition, lead to warmer weather everywhere all the time, just as it will not necessarily result in wetter or drier conditions everywhere. Intricate feedback loops are being discovered that render very accurate predictions almost impossible. Likewise, the impacts on plants and animals can be surprising.

Consider the timing of spring green-up. Scientists at the National Oceanic and Atmospheric Administration using satellite imagery have shown that over the past twenty-five years spring has been arriving progressively earlier in the northern United States. In fact, spring has been arriving there about a day earlier every three years. However, in the southern United States just the opposite is happening. Here spring is arriving later by about a day every seven years. So what's the story here? Aren't we experiencing global warming too? The difference is tied to plant physiology. Most plants need a fixed amount of cool weather before they can fully respond to the first spring warmth. This is not an issue in the longer, colder winters of the North. In the South, though, plants are not receiving the requisite amount of cool weather in order for buds to burst at the first warm snap. It takes a longer period of warm weather to bring them out of dormancy.

If there is a message here, it is that delaying action to address global warming on the basis of not yet having enough data to predict the future is not a viable alternative. We will likely never have the degree of accuracy required by some naysayers.

A Towering Issue

Cell phones—they have become an artifact of our society no less than clay cooking balls were for the prehistoric Poverty Point culture. They are, however, much less innocuous from an environmental perspective. The problem lies with the communication towers that relay our daily chat-abouts and that we naively ignore. Considerable research indicates that as many as 5 million birds are killed in collisions with communication towers in the United States each year. Thousands have been killed in one night under just one tower during certain weather conditions. Tower-related deaths have been reported for 350 species of migratory songbirds, and some, such as thrushes, warblers, and vireos, appear to be disproportionately vulnerable.

The phenomenon is not new, as television broadcasting towers have long been known to pose a danger to migrating birds. The threat has recently increased exponentially with the explosive growth in the number of towers associated with the cell phone industry. The Federal Aviation Administration estimates that there are now eighty thousand towers in the United States, with five thousand new ones being added annually. Hundreds are scattered across Louisiana, including mega-towers taller than 800 feet. While driving through the state one is rarely out of sight of a tower for long.

Birds are killed at communication towers by two different means. Birds flying in poor visibility don't see the structure in time to avoid it, especially fast-flying birds like waterfowl and shorebirds. Towers that are lighted at night for aviation safety may help reduce bird collisions caused by poor visibility, but lights induce a second cause of mortality. During periods of low cloud ceilings or foggy conditions, lights on a tower refract off water particles in the air, creating an illuminated area around the tower. Migrating birds lose their stellar and landscape cues for nocturnal migration in these conditions. When passing the lighted area, birds experience the increased visibility around the tower as their strongest cue for navigation and tend to remain in the lighted space near the tower. The lights do not seem to attract birds but tend to hold those that pass within the illuminated vicinity. Mortality occurs when they collide with the structure, its guy wires, or other migrating birds, as more and more passing birds become disoriented and pack into the small, lighted space.

Measures have been identified that will minimize the impact of tower-related bird kills. Lighting design, tower placement and height, and number of supporting structures are important. Communication equipment should be located on existing towers when possible. New towers should be constructed to accommodate multiple users.

Two thousand years from now archaeologists may ponder the cell phones in our landfills and the ruins of towers that dot the landscape. If we act responsibly today, their contemplation can be interrupted by birdsong.

Recycle

Good examples can be discovered in the most unexpected places. Catch a plane to the Central American country of Costa Rica and land at the capital of San Jose. Get out of town quick because the traffic can be horrendous even with gas at five dollars a gallon. Take a bus northeastward through the vast and mountainous Braulio Carrillo National Park, where crashed airplanes have gone undiscovered for weeks. Continue for 50 miles until the hard-surfaced road forks and becomes a narrow cobbled lane. Bounce over this bone-jarring side road for two hours through corporate banana plantations and steadily increasing heat and humidity until it ends on the bank of a roiling, muddy river. Here the tropical rain forests of a real jungle begin. Hire a rustic boat and motor north through a myriad of canals, rivers, and lagoons for 27 more miles, while flocks of parrots squawk overhead and howler monkeys scream from somewhere inside the adjacent walls of chlorophyll. Finally, pull ashore at the small native village of Tortuguero. Located on the edge of a national park with the same name, the hamlet of Tortuguero sits astraddle a narrow spit of land bound on the east by the Caribbean Sea and on the west by the river. The people are dark-skinned and speak with a rich Jamaican patois. They thrive in an aquatic habitat where children are not allowed to swim in the ocean because of the sharks nor in the river because of crocodiles. The beaches of the area are best known for harboring the largest concentration of nesting green sea turtles in the Western Hemisphere. Walk down the wide sidewalk that serves as the main street of this roadless village until you pass the one-room police station. There in the heart of the settlement one can view a genuine "good example." Four freshly painted containers under a small kiosk are labeled in Spanish for "paper," "plastic," "aluminum," and "glass." How is it that recycling can occur in the remotest jungle of a small, relatively poor, third world country and not in the bayou cities and towns across most of Louisiana?

Resource Exploitation I

Humans have consumed the Earth's plants, animals, water, air, soil, rocks, and minerals since day one. For tens of thousands of years people used the planet's natural resources for the basic necessities of food, water, and shelter. Life was simple even if it was difficult, and the global impact of humans on Earth's resources was generally not significant. Beginning about five thousand years ago social structures evolved in a direction that yielded extracurricular artifacts like the Egyptian pyramids, the Great Wall of China, and the fantastic cathedrals of Europe. Though impressive, these relics are basically big, isolated piles of rocks, the collection of which had little other than local impacts on the environment. It has only been in the last 150 years that humans began to use natural resources at a globally unsustainable rate. Today, two linked factors drive a consciousless engine of exploitation. The first is the exponential growth in numbers of people who crawl, walk, swim, eat, drink, and otherwise consume on, over, and in this planet. The second is the recently attained, technology-laden, wealth-dependent ability of humans to acquire, for lack of a better word—STUFF. Like our own bodies, all STUFF is of this Earth. Nothing came from thin air—and even air is of this realm—and excepting meteorite paperweights all STUFF was once some form of earthly plant, animal, water, soil, rock, and the like.

One example of our STUFF is the automobile. The world contains more than 520 million cars, with 215 million of them in the United States. As consumers with the inherent need to perpetuate our genes, we should at least seriously consider where this STUFF comes from. The steel, aluminum, glass, plastic, and rubber in the typical car did not come from outer space, at least not in the last several billion years. Imagine a hole in the Earth that could accommodate 520 million cars. Cumulatively, this void exists within our bank of natural resources. Now imagine other craters that would hold the rest of the STUFF in the world, things like refrigerators, locomotives, trophy houses, cruise ships, and computers. This perspective should convince us that at our current pace of consumption either our STUFF or the chasms they create will eventually overwhelm our natural resources.

Resource Exploitation II

Having had the opportunity to work up close and in person with the oil and gas industry forty years ago and again recently, I can say without a doubt that it is a different ball game now in terms of sensitivity to environmental concerns. With the enactment of a host of environmental regulations, mainly to prevent the widespread devastation caused by the industry in times past, there is now almost a sense of paranoia to mitigate negative impacts and the resulting bad press. This is good even if the motive is suspect. For years the destruction of many thousands of acres of habitat and its associated wildlife via pollution, the extraction practices, and product transportation and processing was of no serious concern to the industry. In Louisiana the productivity of once pristine marshes was lost forever as miles of dredged access canals allowed saltwater intrusion into freshwater ecosystems. Inland, saltwater was an unwanted byproduct of shallow oil and gas wells. It was routinely dumped into the adjacent terrestrial habitat or water body, resulting in long-term sterilization. Contaminants in drilling fluids and heavy metals like mercury used in metering systems were often abandoned in place. Today, advances in technology, regulatory enforcement, and, most significant, public outcry have reduced many of these kinds of impacts to tolerable levels. The greatest environmental harm now is not pollution but the fragmentation and direct loss of habitat with its resulting consequences. For example, the recent surge in pipeline construction in Louisiana has eliminated thousands of acres of natural habitat for generations to come. The impacts go beyond the actual pipeline rights-of-way. The corridors serve as routes of invasion for undesirable species such as Chinese tallow. Thousands of large timber mats made from hardwood trees are used during construction, thus impacting a scarce resource far from the pipeline. Like pollution, these impacts can also be mitigated, but only by an educated and proactive citizenry.

Resource Exploitation III

Dictionaries define the word "finite" as "having bounds; limited; existing, persisting, or enduring for a set time only; impermanent; not infinite." Doesn't seem to be a gray area to me. Most everyone agrees that the amount of crude oil remaining in this planet is finite, yet we behave as though the tap will never run dry. Americans in particular have an insatiable thirst for gasoline, consuming 178 million gallons every day in order to drive 2.5 trillion miles per year. Numbers this large leak out of my cerebrum as fast as they are poured in, but the idea of this mileage being equivalent to 14,000 round trips to the sun convinces me that things have changed since the horse and buggy days. How can this gallop be maintained?

It takes a while to make crude oil—about 300 million years is a good average. We're not talking Sparta Aquifer here with a recharge rate measured in decades. Most oil was produced in the Carboniferous period, when the land was covered with swamps harboring an abundance of prolific vegetation and the oceans were rich in algae-like organisms. As the plants died they settled to the bottom of swamps and oceans to be deeply covered by various sediments over eons. Increasing pressure from the overburden and heat from the inner Earth cooked this gumbo of tree ferns, cycads, and diatoms until it eventually became a roué of oils with varying viscosity, natural gas, and perhaps coal. The formation of crude oil occurred at a unique time in the history of our planet under specific climatic, biological, and geological conditions. These conditions no longer exist. The fire under the gumbo pot has gone out. For the sake of our grandchildren, we'd best go on a crude oil diet while searching for a new source of kindling.

Resource Exploitation IV

History is replete with examples of the exploitation of natural resources resulting in fabulous wealth for a few individuals, often to the detriment of others. North American cases include John Jacob Astor, America's first multimillionaire, who based his fortune on fur trading, but was accused of cutthroat, monopolistic business practices. Andrew Carnegie became the richest man in the world via the steel industry with a foundation in the mining of iron ore. He was criticized for brutal labor practices. Timber barons are noteworthy for their "cut out and get out" policies of leaving the local populace destitute after complete removal of forest resources. Such illustrations are not all in the distant past. Not long ago the new CEO of a large timber company based in the Pacific Northwest, but with considerable holdings in Louisiana, announced a policy that amounted to slash-and-burn of their old-growth forest tracts by telling his employees that he believed in the Golden Rule—that is, "Whoever has the gold rules." In northwest Louisiana vast deposits of natural gas are being developed with the advent of new technology in a geologic formation known as the Haynesville Shale. Large landowners are becoming instant millionaires. One owner of less than 400 acres reportedly received a lease check of $18 million even before his 23 percent share of royalties began. The relatively few people fortunate enough to own large tracts of land there live in a parish where 30 percent of the residents live below the poverty level.

History also reveals that those who exploit natural resources, unethically or otherwise, are capable of selfless, social benevolence. After retiring, Aster became a patron of culture, building libraries and supporting the poor. In his old age Carnegie had a change of heart and declared, "The man who dies rich dies disgraced" and proceeded to give away his fortune. The verdict is still out for those yet in the harvesting arena.

Moths and Mulberries

The Civil War may be indirectly implicated in the continuing devastation of American forests by an army of insects proven to be
less stoppable than those of the Union or Confederacy. When cotton from Louisiana and other southern states became unavailable
in the North, Leopold Trouvelot, a Boston naturalist, accelerated
his research into producing a viable silkworm for the northern tex-
tile industry. It led to his infamous, late 1860s importation of gypsy
moth eggs from France in an effort to cross breed them with silkworm moths. Because the two insects were only distantly related,
they could not interbreed and the experiment failed. Either accidentally or intentionally, gypsy moth caterpillars were soon released
into a hospitable environment and became the gypsy moth plague
that causes damages valued in the hundreds of millions of dollars to
North American forests each year.

The intended host plants for the silkworm caterpillars were native red mulberry and white mulberry, an introduced Asian species that serves as the foundation of silk production in other parts
of the world. Promoted by the U.S. government, white mulberry was
widely planted before the Civil War in an unsuccessful effort to develop a domestic silk industry. Like the gypsy moth, white mulberry
has become an invasive species with negative ecological impacts
on natural areas. It hybridizes with and transmits disease harmful to red mulberry and displaces other native vegetation. So, instead of a viable silk industry in the United States we ended up with
economic, aesthetic, and ecological havoc in the forests of eastern
North America as lingering impacts of a civil war soon to mark its
sesquicentennial. For the environment, too, war is hell.

Thirst

As residents of a state with abundant precipitation, miles of rivers
and bayous, and thousands of acres of lakes, we rarely give a second thought to our source of drinking water. In the mid-1970s,

I lived for a while on an old homestead in the shadow of Driskill Mountain, the highest elevation in Louisiana. There the sole source of water for drinking, cooking, washing, and life in general was a small spring behind the dog-trot house. For many people in the hill parishes, shallow, hand-dug wells and springs provided water until subsidized community water systems, which relied on deep, bored wells, were developed. Dependable springs were a treasured resource on any property.

Where springs exist, they are conduits in the water cycle. Rainfall seeps underground by percolating through tiny spaces between soil particles and is stored in porous sands or rocks. Spring water is considered young water because it usually has fallen as rain months or a few years before. The exact land surface where water seeps underground and contributes to a specific spring is called that spring's recharge basin. In areas where soils are mostly heavy clays, such as the delta parishes, rainfall doesn't soak in as fast and runoff into lakes, rivers, and other wetlands is a more natural process. For this reason springs are rare in the alluvial flatlands of Louisiana, and most early settlers in those regions depended on cisterns for a drinking water supply.

Springs were very common in the hill parishes. They formed where groundwater was forced up to the surface as a result of differences in slope in the shallow aquifers. As rain falls and percolates underground, pressure is exerted on water already in the aquifer and forces some out through natural openings. The unique habitat created by springs attracts unusual animals as well as humans. A spring in Jackson Parish is home to five species of caddisflies found nowhere else. Others harbor rare salamanders.

A spring near my house in Union Parish once served the household needs of several families and the boiler of my great-grandfather's small, steam-powered cotton gin. Fresh milk was kept in the cold spring box, which also served as the home of a goggle-eye bream specially chosen to control mosquito larvae and other insects in the clear pool of water.

This spring and most others in the region are mere remnants of their once flowing glory. Many have disappeared completely as a result of depleted water tables and developed or disturbed recharge basins. Shallow aquifers are particularly susceptible to contamination from fertilizers, pesticides, and other pollutants, making even the survivors unsafe as drinking water. Had we been more astute, their unheralded loss could have been an early warning of the problems that we face today with the critical deep aquifers that fill our iced tea glasses of mid-summer with increasing reluctance.

Lacassine Visit

More than twenty-five years ago I was a fledgling wildlife biologist on one of the most spectacular wildlife areas in the country. Lacassine National Wildlife Refuge, comprising almost 35,000 acres in southwest Louisiana, contained some of the most beautiful, biologically rich, freshwater marshes anywhere. More than 3,000 acres of natural marsh is an officially designated Wilderness Area, but the jewel of this national treasure was Lacassine Pool, a 16,000-acre impounded and managed marsh. During my tenure one could stand on the levee of Lacassine Pool on a winter evening at sunset and observe a wildlife extravaganza unique in North America, as up to a million waterfowl flew in and out of the pool on foraging missions. The sensory overload of such scenes evoked in me the idea that this is what pristine nature looked like before humans altered our world for good.

Not long ago I had an opportunity to revisit Lacassine National Wildlife Refuge and ponder a quarter century's worth of changes. Nothing in the natural world is static, and some of us who occasionally long for the good old days should know better. We realize that natural phenomena, such as plant succession and wildlife population fluctuations, are normal. Even the changes caused by Hurricanes Rita and Gustav are natural in the long-term scheme of things. It's the negative human-induced changes, intentional or otherwise, that stick in the craw.

Gliding down the bayou that transects the refuge in a bateau, the first thing that jumps out at me is that the marshes look weedier. Chinese tallow trees, which have been here a long time, are now more abundant and have been joined by new insidious invasive species like common salvinia and witch grass. They take the places of native plants and are up to no good. A 100-foot oil-drilling derrick rises up out of the pool, something that formerly never occurred in winter when ducks and geese were a priority in an inviolate sanctuary. Instead of a million waterfowl, we see a few thousand, which has been the norm for a while, and yet the hunting seasons are longer and the bag limits more liberal now. The birds are not somewhere else, as some believe.

We saw positive changes, too: fewer exotic nutria, more roseate spoonbills, a prairie restoration project, and a new wildlife drive. Overall, though, the biological integrity of Lacassine has declined both because of and in spite of human actions—an ominous dividend of the political, economic, and social trends in recent years.

Tree Pathogens

North American trees are being assaulted by foreign pathogens no less virulent than many thought swine flu to be in human populations. The potential for widespread devastation is just as real and has historical precedence similar to the flu pandemic of 1918. By 1940, chestnut blight had eliminated the 4 billion American chestnut trees in the United States. This keystone species of eastern forests simply vanished as a functioning component of ecosystems once defined by its presence.

In recent years new waves of tree diseases have turned up on our shores. Most are from Asia because similar species of trees grow in that region. Over thousands of years they have developed a resistance to local diseases. When the pathogens arrive in America, related but vulnerable species have no such resistance.

Examples include sudden oak death caused by a fungus that travels in raindrops. It arrived in California in the 1990s and began killing oaks of all species. It's heading toward Louisiana now.

Hemlock wooly adelgid, an Asian insect that feeds on the plant fluids of hemlocks and leaves a tell-tale egg sack that resembles a cotton swab attached to the bottom of needles, has decimated hemlocks in the eastern United States. The once giant specimens in the Great Smoky Mountain National Park now stand as lifeless snags. Dogwood anthracnose arrived on both coasts in the 1970s. It has reached Louisiana and routinely kills our native dogwoods. The fungus produces cankers usually starting on the lower limbs and moves upward to envelop the tree in disease. Forest conservationists are most concerned today about an insect that was discovered in Michigan in 2002. The emerald ash borer likely came from Asia in the lumber of wooden packing crates and has moved as far south as Missouri, killing tens of millions of trees in its wake. Quarantines are in place on ash from infected states.

Trees have been subjected to diseases for millennia. The difference today is the rate at which our native forests are being exposed to new pathogens. Just as European diseases like smallpox extinguished Native American cultures, the diversity and richness of our forests are similarly threatened. To date, most of these exotic diseases have proven unstoppable. Like swine flu in humans, they are an example of the unpredictable nature of nature.

Light Pollution

Someone is stealing the stars. The odds are great that the number of stars visible from your backyard on a clear night is considerably less than ten years ago. A view of the night sky a century back would convince you of this insidious theft. Under the darkest skies left on our planet, about 2,500 stars are visible to the naked eye. In the suburbs only 10 percent of these can be seen, and city folks are lucky to see a few dozen.

The culprit is light pollution. Light from millions of sources washes away the brilliance of night skies. The loss is more than aesthetic. Estimates of wasted energy due to inefficient lighting approach $1 billion annually. Astronomers gnash their teeth as light from encroaching cities renders observatories impotent for important projects. Natural ecological processes are affected too. Migrating songbirds are thrown off course. Newly hatched sea turtles will crawl toward bright stadium lights instead of the safety of the ocean. Artificial lights will even cause trees to delay shedding their leaves.

Terminology specific to the issue has evolved. "Light trespass" is light that is present where it is not wanted or needed. It may be your neighbor's overbearing security light that shines in your bedroom window. It may be your light shining in his. "Uplight" is light that goes directly up into the night sky and serves no useful purpose. There is even an organization, the International Dark-Sky Association, whose objective is to educate people about light pollution.

Experts on the subject agree that light pollution can be reduced economically without impacting convenience and security. The key is matching the appropriate lighting fixtures to the situation. Low-pressure sodium lights are good substitutes for glaring street and security lights. Good lighting is also well shielded to direct the light where it is needed and prevent uplight.

Unlike other types of pollution, light pollution has not proven to be a serious human health problem, at least in the physical sense. Psychologically, it may move our civilization one step farther from the natural world that nourished our ancestors. As children, our grandparents' grandparents were more in tune with the rhythm of the night sky than we are today. Some measures of progress are subjective.

6
JUNCTION
The Act or Process of Joining to Become One in Harmony

> Friends of Black Bayou [a nationally recognized Louisiana conserva-
> tion advocacy group] Fall Celebration event, 10th Anniversary; 2,000 in
> attendance.
>
> —KO Field Diary, 13 October 2007

> Good article today in "New York Times" about our efforts to remove
> levees on Mollicy Unit of Upper Ouachita NWR and restore natural
> hydrology to 16,000 acres of bottomland hardwood habitat.
>
> —KO Field Diary, 20 June 2009

Attitudes

Since the founding of America, attitudes toward nature have
changed, and they continue to do so. Early pioneers maintained a
European mindset, considering nature an entity to be conquered,
civilized, and rid of competing wild beasts as necessary. The theory
of manifest destiny reflected a theological belief that settlers were
divinely appointed to "use" the Earth for the enhancement of civilization, no holds barred. Such attitudes eventually led to the decimation of Native Americans and the extinction or near extinction of
several animals.

Barely tempering this prevalent thought in the 1830s and 1840s, the poets Lord Byron and William Wordsworth and artists associated with the Hudson River school of painting championed a "romantic attitude" in which nature was promoted for its aesthetic values. With his publication of *Walden* in 1854 Henry David Thoreau argued not for sentimentality but for the wisdom of seeking God in nature. His position contained the embryonic thoughts of what later became known as the "preservation attitude."

The aftermath of the Civil War left a country wallowing in reconstruction and soon to begin a broad-based assault on the wild flora and fauna of North America. Two factors fueled the destruction: new technology and the rush to rebuild the South. Both resulted in the near total elimination of virgin forests in the eastern United States. Wildlife was not spared. Within two generations of the war, regional and global extinctions of such species as eastern elk, red wolves, ivory-billed woodpeckers, whooping cranes, Carolina parakeets, and passenger pigeons occurred. The saga of American bison, plume-bearing wading birds, East Coast fisheries, and waterfowl is well documented. Wildlife populations plunged because of loss of habitat and direct overexploitation such as market hunting that accelerated with the development of railroads and refrigeration. The biblical mandate to subdue the Earth was pursued in earnest and few protested.

The tide began to turn after misuse of natural resources of the eastern United States peaked, and large-scale clearing of western forests along with the slaughter of millions of bison began. Enough influential people valued a frontier of some sort that the preservationist movement was born to set aside and protect at least a small part of it. The era of conservation followed with intensive efforts to manage and enhance the remaining populations of plants and animals, especially those with monetary or sporting value. The present era of conservation biology attempts to address flora and fauna on a broad landscape scale encompassing all ecosystems on the planet. In the history of humans and nature, the challenges are unprecedented.

Demography

Demography is the statistical study of human populations. Little is known about the demographics of Native Americans in northeast Louisiana prior to European settlement. Abundant archaeological evidence indicates that Indians frequented this area for at least ten thousand years, but that their population levels fluctuated over time. Large, complex sites, such as Poverty Point and Watson Brake, supported many people, but their cultures vanished thousands of years before Europeans became established. One thing is known for sure: when French Canadian trappers entered the area in the mid-1700s, the only folks around were small, itinerant bands of natives.

European settlement progressed from the eastern seaboard. Many northeast Louisiana settlers were second- or third-generation Americans born in Georgia, Alabama, and Mississippi. They were farmers and cleared the fertile lands along the banks of the major waterways. By 1860, just prior to the Civil War, four times as many people lived in East and West Carroll parishes as in Ouachita Parish, today's metropolitan area. The population of Tensas Parish was three times that of Ouachita, and even the hill country of Union Parish supported twice the population of Ouachita. The state as a whole was still rural even though one out of four citizens lived in New Orleans.

In spite of the economic devastation of the Civil War, the population continued to increase but the pattern changed. By 1900, Ouachita Parish was the most populated. The most recent census in 2000 indicated that Ouachita Parish now has nearly seven times the people of East and West Carroll parishes. While Tensas Parish has less than half the population it had in 1860, Ouachita's is now 30 times greater than it was then. Madison Parish lost 22 percent of its population just between 1980 and 1990.

The reasons for the demographic shifts through time are complex, but natural resources were always the driving force behind human settlement patterns and subsequent migrations. People were first attracted to abundant wildlife for subsistence. Later immigrants sought fertile soils for agriculture and waterways for transportation. The evolution of technology allowed alteration of the environment on a landscape scale and the exploitation of natural resources heretofore unavailable. Most people lived where the resources were located—on the farms or in the small sawmill towns. In most situations, resources were used up faster than they could regenerate—market hunting depleted wildlife populations, the virgin forests were cut, soils in the hill parishes were impoverished, natural gas production plummeted. People moved on.

Continuing technological advances reduce the need of onsite manual labor and allow people to concentrate in cities farther from the resources. People can now efficiently travel out to the resources or bring them in to cities and towns for processing and utilization. Demographic vacillations in northeast Louisiana will continue as they have for thousands of years. The direction will depend on the status of our natural resources and our ability to use them wisely.

Soldiers

I am watching soldiers. They are on patrol now. Two of them, they are armed with a long-handled, fine-meshed dip net and a white plastic bucket. Their assignment for the morning is to search for and capture a dozen each of the tribes *Gambusia* and *Fundulus*. They are excited about their mission; here even the word "fervent" is not an overstatement. The soldiers are 9 and 13 years old but quick to announce that in just over a month they will be 10 and 14, respectively. Wearing a pork pie hat and baggy, camo-patterned shorts, the youngest boy is a prime age for the job, and his pony-tailed brother is first-rate also, not yet having succumbed to the pseudo-insight of adolescence. They are having an experience increasingly rare among children in this country: a pure, unadulterated encounter with nature. No electronics are involved. The fantasies are their own, not those of a software developer or a TV producer. Creativity is limited only by the size of the pond around which they stalk. Boys so engaged are immune to the pervasive sauna of a mid-summer Louisiana morning. All senses are focused on the darting minnows that more often than not evade the flailing dip net. A yearling bass competes with the boys. Without realizing it they are learning the basics of human survival: the minnows are concentrated in the shelter of natural, aquatic vegetation; food pyramids surface; minnows can live in the pond because the water is unpolluted; the pond exists because no one pumped it dry for agriculture or filled it to support a shopping mall or trophy house. Below the threshold of awareness, the boys begin to formulate the natural cornerstone of wisdom—an environmental conscience. As such knowledge is necessary to sustain mankind in the future, deploy your soldiers to the ponds.

Field Guides

Your Louisiana natural history library is incomplete without a set of good field guides. Field guides are specialized books used to identify natural objects in the environment. An individual guide usually covers a particular group, such as birds, rocks and minerals, trees, or seashells. Their scope may be as general as "Insects of North America" or as focused as "Butterflies of the Southern Gulf Coast." They are called field guides because they are small, compact, and easy to carry on a hike or other outdoor excursion.

Most modern field guides use photographs or color plates to aid in the identification of the unknown object. It is often a matter of just leafing through the book until you find a likeness of your item. However, some objects, small female songbirds, for instance, are very similar in appearance and not easily identified. For this reason, it is important to read the often-ignored section found in every field guide that is usually titled "How to Use This Book." Here the logic behind the arrangement of the objects is explained. Some are depicted according to color or size, others according to taxonomic relationship, and still others according to where they are found. Specific hints offer guidance that assist the identification of difficult objects, often by the process of elimination. Small gray birds, for instance, may be separated by those with and without white wing bars. Yellow wildflowers can be divided into those with opposite or alternate leaves on their stems. Continuing in this manner, identification can usually be determined.

Range maps are an important part of most field guides. They indicate where specific objects are normally found and help to eliminate confusing possibilities. If you are interested in determining what type of mole is burrowing in your okra patch, look in a field guide to mammals. Four of the five kinds of listed moles are found nowhere near Louisiana, thus a quick and easy ID. Many similar plants and animals can be identified by their habitat. If you have decided that the native tree in your back yard is either a green ash or a white ash according to photos in a field guide, check the text describing their natural habitats. Green ash usually grows in areas that were once bottomland hardwoods, and white ash grows in upland areas.

Among others, there are field guides to fish, fossils, mushrooms, and roadkills. Available at most bookstores, they make excellent gifts. They can be effective tools to stimulate curiosity and a desire to learn in grandchildren and grandparents alike. To appreciate the diversity of natural history in bayou country, it always helps to put a name on it.

Going, Going, . . .

Congress will soon be struggling again with reauthorization of the Endangered Species Act of 1973. Opponents of the act claim that recent interpretations of the law have gone too far and are depriving citizens of basic rights, such as those associated with land ownership. Proponents say the act is not stringent enough in protecting rare and endangered species and cite examples of continuously declining populations. A rational solution probably lies midway between the most extreme viewpoints.

Since Europeans arrived in what is now Louisiana, several species once found here have become globally extinct, and others have been extirpated from the state but survive in other locations. Millions of passenger pigeons migrated to Louisiana before market hunting eliminated them forever. The last known record for Louisiana was of a few individuals observed near Mer Rouge in Morehouse Parish in the winter of 1902–3. Carolina parakeets, with their bright green bodies and yellow and orange heads, met a similar fate because of their fondness for farmers' grain fields. The last known viable population of ivory-billed woodpeckers occurred on the Singer tract in Madison Parish, now a part of the Tensas River National Wildlife Refuge. Massive landclearing in the lower Mississippi Valley eliminated their habitat. The last definitive, widely accepted report of the bird is of a lone female that lingered in this area in the spring of 1943, after the felling of a tree that contained a nest and eggs the same year. Bison, or buffalo, were common in Louisiana and were often reported by early French explorers. Boeuf River, Boeuf Lake, and Bayou Boeuf are named after bison. In March 1700, Jean Baptista Le Moyne, Sieur de Bienville, reported killing bison near present-day Winnsboro and later in what is now Winn Parish. The last record of the presence of bison in the state was of one shot in 1803 near Fort Miro, the present site of Monroe. Even elk were once found in Louisiana, although they were likely never very common. A bull was reported shot in December 1842 near Mound in Madison Parish.

Because large predators hindered settlers' attempts to raise livestock, species such as the cougar and red wolf were relentlessly pursued throughout the nineteenth and early twentieth centuries. The red wolf is considered extinct in Louisiana, and the few substantiated cougar reports in recent years seem to be widely roaming individuals originating far outside the state's boundaries.

Not all of the news is gloomy, and real success stories exist within Louisiana. Alligators are a prime example. With careful management, alligator populations once decimated by poaching now thrive. The brown pelican, Louisiana's state bird, disappeared from the state and was successfully reestablished through restocking. The number of bald eagles in the state continues to increase as a result of stringent protection and the ban on the chemical DDT. Challenges still remain with species such as the Louisiana black bear, red-cockaded woodpecker, and whooping crane, but it is important to remember that whatever your views on the Endangered Species Act, the record has shown that humans have the ingenuity and means to save most species, if they have the will.

Living It Up

I find it hard to believe that so many people are terrified of hummingbirds, cardinals, box turtles, and swallowtail butterflies. This unreasonable behavior is further characterized by a deep animosity toward wildflowers and indeed any plant that is not mass produced in commercial nurseries and touted in the latest home and garden magazines. Why else would people go to such extremes to sterilize the environments around their homes and eliminate every square foot of habitat for these beautiful and fascinating native flora and fauna? Irony lies in the fact that many of these same people love to visit parks and similar areas to enjoy the same natural features and creatures that they spend tremendous effort and money to destroy in their own yards. It is probably not a stretch to assume that an inverse relationship exists between the annual sales of lawnmowers in the United States (10 million at last count) and the number of fern beds and butterfly gardens.

I would like to believe that this situation occurs simply because people just don't think about it. They grew up in sterile mowed yards, their neighbors had sterile mowed yards, and the house they live in had a sterile mowed yard when they bought it. It doesn't have to be. It's not an either/or situation. Lawns do have their place and are actually desirable in some locations, if they are part of a landscape that includes natural elements. Any yard, anywhere, can accommodate wildlife habitat. It may be a butterfly garden on the narrow strip between a sidewalk and brick wall of an apartment, a lily pond, wild iris and bird feeders in the back yard of a subdivision home, or a 5-acre arboretum at a country estate. If we provide it, they will come—and enrich our lives in the process.

Fences

About a hundred yards north of my house in the dense woods, the remnants of an old fence can be seen running north-south over a sandy-clay hill on the edge of the D'Arbonne Swamp. The forest looks the same on both sides of the rusty wire now, but it once enclosed a 10-acre field where my father chopped cotton as a teenager. When boll weevils, armyworms, and worn-out soil forced the Union Parish hill-country cotton farmers to seek work in paper mills, in chemical plants, and on pipelines, the field reverted to forest through natural plant succession. The timber on it has been cut at least twice since the Great Depression, the last time about 1988. I moved next door to the property a couple of years later and remember finding grog-tempered potshards in the loader sets. It occurred to me at the time that this evidence, along with a few chert artifacts, was the only indication that hundreds of generations of humans had lived here long before white settlers of European descent began offloading up the river at Alabama Landing. For better and worse, the lingering and continuing changes to the local natural world can be attributed to the offspring of these new people. We regal in the better but are blissfully ignorant of the worse. The biological sterility of commercial pine monoculture has swept clean the rich biodiversity of historical upland hardwood forests. Even the once abundant free-flowing springs that nurtured Native Americans and settlers alike have disappeared into plunging aquifers, collateral damage of unquenchable local industries.

The cotton field is gone now, and I often wonder what this area will look like in a hundred (or even a half dozen) generations. Nature is remarkably resilient. A forest *can* restore itself with biodiversity if demand for single-species pulpwood is assuaged, and aquifers *can* be replenished, but only when the fence that restrains thoughtful consumerism rusts into the shadow of time.

Bird Feeder

Nothing can brighten a gray winter day faster than a splash of crimson cardinals or goldfinches gathered at a window-side bird feeder. The popularity of bird feeding continues to grow, and a recent report estimates that 63 million Americans provide food for wild birds, spending more than $2.5 billion on birdseed and feeding supplies each year.

Winter is often a stressful time for species that are year-round residents in Louisiana, such as the cardinal, and those like goldfinches, which normally only spend the winter here. Days are short and cool; nights are long and often cold. As winter progresses,

berries and seeds become scarce and insects disappear. At this time feeders can be helpful in carrying local birds through to spring in good nutritional condition.

Most grocery, hardware, and discount stores offer a wide variety of birdseeds, and it's important to know the types that are actually preferred by birds. Different kinds of birds prefer different types of seeds, but sunflower seeds attract the greatest number of species. Several studies have shown that this high-energy food is by far the favorite of most species that visit feeders. Black oil sunflower is best. Striped sunflower seeds have larger, thicker seed coats, making them difficult for small birds to handle and crack. The bags of mixed seeds containing sunflower, millet, milo, oats, and others are often not good bargains because birds kick out and waste many of the smaller seeds to get to the sunflowers. Some types of birds, however, do prefer seeds other than sunflowers. Doves like white millet seeds, jays and quail like corn, and finches often prefer thistle seeds. You can even prepare your own bird food by saving and drying seeds of squash, gourds, melons, and cucumbers. Chop the larger ones in a food processor.

Insect-eating birds like nuthatches and woodpeckers can be attracted with beef suet or peanut butter. In Louisiana, peanut butter works best because suet quickly becomes rancid on a warm day. Special wire containers to hold these items can be bought, or they can be offered in a plastic mesh bag or in holes drilled into a log. Other birds that don't normally come to feeders because they don't prefer seeds, such as robins, thrushes, bluebirds, mockingbirds, and cedar waxwings, often respond to an offering of fruit. Try raisins and fresh orange and apple slices.

Some responsibilities come with the pleasures of bird feeding. Congregating birds makes them more susceptible to disease and predation from cats. Be sure to store seeds in tight, waterproof containers safe from rodents. Don't feed moldy or mildewed seeds. Thoroughly wash and sterilize bird feeders with a weak solution of bleach at the beginning of each feeding season. Domestic cats kill millions of songbirds. Make sure all feeders are cat proof.

Winter feeding of birds is addictive and often leads to year-round habits. A host of dedicated conservationists began their informal schooling in natural history via a simple bird feeder. Enroll now.

Refugees

From the window of my second-floor office at Black Bayou Lake National Wildlife Refuge I can see about two hundred of them, mostly fifth graders but also a smidgin' of teachers, mommas, and daddies. They were recently deposited here by four large, yellow-orange school buses that now sit idle on the parking lot in stark contrast to the green of this place. The little people represent well the diversity of our local population. Light skin and dark skin, long hair and short hair, round and thin. Unlike the adults who wear t-shirts with "Race for the Cure" and "Mardi Gras" logos, the children are divided into groups of brown shirts, green shirts, yellow shirts, blue shirts, and red shirts. They travel in these homogenous packs and, from above, resemble giant amoebas sprouting cilia of squirming arms and legs from every surface. Woe be it to the individual cell who tries to abandon its organism of color. If a blue shirt person should break away and drift toward the red shirt creature, it triggers a howl of screeches and fluttering cilia that quickly sucks the repentant truant back into its proper place. Peer pressure is more potent than Maxwell's theory of electromagnetism here.

The groups all have missions. One bunch flails the display pond with dip nets like prehistoric predators and races for the microscopes with their dripping prey of water bugs and other things tiny that wiggle. Another unit, armed with field guides, tramps the nature trail and argues among themselves about the identity of a woodpecker—hairy or downy, and the tree that it is pecking—oak or elm. A scavenger hunt sends another cluster foraging for feathers, lichen, and turtle shells—all things organic.

The hope is that there remains in the vacuoles of their protoplasm space that is not entirely filled with video games, television, and contests played out on concrete—space that will harbor the message of the day captured amoeba-like by simple absorption.

Coming Back

Pendulum-like, her right arm swept downward in an arc inches behind the spinning steel disc and back up to the hopper. Each cycle took less than two seconds, and each changed the world perhaps for generations. The biologist was riding a tractor-drawn implement, planting hardwood seedlings in a vast agricultural field in the middle of a swamp. The area had recently been acquired as a federal wildlife refuge, and mitigating past human arrogance was a priority. As the coulter sliced open the earth, the momentum of the tractor literally pulled the seedling from her hand as it was placed in the incision. Trailing packing wheels sealed the wound.

The earth received the seedlings hungrily; they were not the most recent genetic mutants of a grass from the Middle East, a mallow from Africa found in blue jeans, or an oily bean from Asia. They were manifestations of nursery-grown wild seeds. Progeny of sporadic glaciers, thousands of floods, and intermittent fires, they had evolved with specific pollinators, microrhizoans, woodpeckers, and boring beetles. They had adapted to survive a host of herbivores and even dendro-chemical attacks by their neighbors. They could not, however, withstand the force of a D-8 bulldozer blade.

The planter was loaded with species native to the swamp. As the field changed slightly in elevation or soil type, the biologist reached for different trees. Cypress and tupelo in the lowest areas, overcup oaks 6 inches up the "hill," then willow oaks and green ash. Diversity—always seeking life-sustaining, system-supporting diversity—sweetgums, pecans, mayhaws, persimmons.

In addition to seedlings, the hoppers were filled with invisible life to come. Their seed spilled into the furrow to sprout parula warblers gleaning insects from Spanish moss in the top of an oak tree. It grew into black bear cubs born in a hollow cypress and largemouth bass spawned on a log during a spring overflow. Fireflies and fungi, crawfish and catfish were planted alongside gobbling wild turkeys and cottonmouths with attitudes.

Trucks delivered bags of new seedlings to the turn-rows. Fresh crews relieved planters during lunch breaks so the tractors could keep running hour after hour and acre after acre. The window of opportunity was small. A rainy spell on hydric soils would end the season's work. The cumulative impact was important and more than biotic. The forest would stabilize and enrich the tons of topsoil that currently washes into the adjacent river. Downstream towns would benefit from the sponge effect of a natural swamp during the ever more frequent "hundred year floods." Filters of a functioning system would trap and break down pollutants.

It is hard to imagine that men once made a cursory examination of this intact swamp and made a boardroom decision to use investors' money to remove the forest for the sake of profitable agriculture. Apparently, rational men didn't, as profits were scarce as screech owl nests in a rice field. Within a few weeks, though, spring rains will erase the chevron-shaped tracks of the tree-planting rig and tractors will till this swamp no more.

EPILOGUE

On April 20, 2010, the Deepwater Horizon rig leased by BP exploded in the Gulf of Mexico 78 miles southeast of Venice, Louisiana, killing 11 workers. The event was the beginning of what President Barack Obama called the "greatest environmental disaster of its kind in our history." Estimates of the amount of crude oil that gushed wildly into the Gulf over more than 3 months were as high as 184 million gallons.

The office of Delta/Breton National Wildlife Refuges is located at the end of the road in Venice on Grand Pass near its confluence with the main stem of the Mississippi River. Within 10 miles of this place in all directions, like capillaries at the terminus of North America's largest drainage, lie Flat Boat Bayou, Lost Dog Bayou, Long Island Bayou, Buras Bayou, Felice Bayou, Sullivan Bayou, Dalton Bayou, Goat Bayou, Dead Man Bayou, Bottle Bayou, Sawdust Bend Bayou, Oil Mine Bayou, Bayou Tambour, Bayou Little Channel, Bayou Dum Barr, Bayou Petit Liard, and Bayou Tony.

> KO Field Diary, 2 May 2010
> 11:30AM—I am writing this from the office of Delta NWR. This second floor room has large windows all around, and I can see the Mississippi River and adjacent waterways filled with commercial boatyards, piers and docks. The weather is harsh, 23 mph sustained winds from the south with gusts to

34. It is pushing northward a menace that has alarmed the world. It is headed this way. A crisis of unknown proportion is imminent.

12:50PM—Four large helicopters, like giant dragonflies, suddenly appeared and began landing 200 yards north of this building at the Chevron shore base. One is a Blackhawk, two behemoths carry U.S. Marine Corps emblems, and a smaller one is labeled United States of America. We received word that the president of the United States would soon arrive in his motorcade, driving down from New Orleans because the weather did not fall within safety parameters required for him to fly in the helicopters. A Coast Guard cutter cruises slowly below my window.

2:20PM—The presidential motorcade arrives, twenty-six vehicles strong, with an armored car, ambulance, and three identical, black SUVs with dark-tinted windows. They pass our office and stop at the Coast Guard station for a news conference.

2:40PM—Another helicopter identical to the smaller one arrives. Redundancy is obvious.

3:30PM—The motorcade returns and stops at the end of our driveway. The president gets out of one of the SUVs and conducts an informal yet orchestrated question and answer session with local commercial fishermen concerned about the permanence of their way of life. He is facing me wearing a dark jacket and blue shirt with an open collar as he gestures to the worried men in white boots.

3:45PM—Everyone remounts as the motorcade turns around and drives the few hundred yards to the heliport. The most powerful man on the planet boards one of the smaller helicopters.

3:58PM—In an extraordinary show of might, the helicopters roar to life and fly north with the wind. It strikes me that they are all the color of oil.

On the following afternoon we eased out of the boat slip at the office in a 29-foot catamaran powered by twin 250 horsepower Suzuki outboards. The big (gasoline-consuming) motors had plenty of reserve as we cruised at 30 knots across the Mississippi River, down another pass, and out into the Gulf of Mexico. In just over an hour we set anchors on the leeward side of North Breton Island and waded ashore. This place knows of the fickleness of permanence in natural systems. The remnant of a much larger island, it once harbored a schoolhouse for the residents before hurricanes reduced it to 40 acres of scrubby mangrove and sand spits. Later, in 1904, Teddy Roosevelt in an executive order declared it the second National Wildlife Refuge in America.

On this day the place was spectacular. Thousands of shorebirds, gulls, and terns whirled and swirled in clouds of wings, some in rigid formation, others beating across the grain intent on life's chore of the moment. Two peregrine falcons slashed through the flocks causing short-lived panic. This vibrant pulsing of activity mesmerized us for a while, making it difficult to remain on-task. Our mission involved another species, one that was once extirpated in Louisiana even though it is the official state bird. Thirteen hundred pairs of brown pelicans, not long removed from the endangered species list, were nesting in the island's mangroves. We had come to assess

Brown Pelican

Bayou Choupique
Bayou Citamon
Bayou Clear
Bayou Close
Bayou Cocodrie
Bayou Conway
Bayou Copasaw
Bayou Corne
Bayou Couba
Bayou Coulee
Bayou Courtableau
Bayou Crab
Bayou Cutoff
Bayou D'Arbonne
Bayou Dan
Bayou Darrow
Bayou de Butte
Bayou de Cade
Bayou de Glaize
Bayou de L'Outre
Bayou de Muse
Bayou de Siard
Bayou Derbonne
Bayou des Allemands
Bayou des Cannes
Bayou des Glaises
Bayou des Saules
Bayou des Sot

Bayou Dorcheat
Bayou Doza
Bayou du Lac
Bayou du Large
Bayou du Portage
Bayou Dum Barr
Bayou Dupont
Bayou Duralde
Bayou Eugene
Bayou Fisher
Bayou Folse
Bayou Fordoche
Bayou Francois
Bayou Funny Louis
Bayou Fusil
Bayou Fusiler
Bayou Galion
Bayou Gauche
Bayou Geneve
Bayou Gerance
Bayou Grand Cane
Bayou Grand Coteau
Bayou Grand Louis
Bayou Grand Marais
Bayou Grande Cherriere
Bayou Gravenburg
Bayou Grosse Tete
Bayou Hopper

Bayou Jack

Bayou Jeansonne

Bayou Joe Marcel

Bayou Johnson

Bayou Jonas

Bayou L'Eau Bleu

Bayou L'Embarras

Bayou L'Ivrogne

Bayou la Cache

Bayou la Nana

Bayou la Rompe

Bayou la Ville

Bayou Lacombe

Bayou Lacombe Tassin

Bayou Lafourche

Bayou Lamourie

Bayou Lassene

Bayou Latenache

Bayou Little Channel

Bayou Little Teche

Bayou Lonfouca

Bayou Luce

Bayou Macon

Bayou Mallet

Bayou Manchac

Bayou Maringouin

Bayou Marron

Bayou Marsh

Bayou Matherne

Bayou Mauvais Bois

Bayou Milhomme

Bayou Milligan

Bayou Misere

Bayou Moreau

Bayou Morengo

Bayou Na Bonchasse

Bayou Natchez

Bayou Natchitoches

Bayou Nezpique

Bayou Paul

Bayou Pays Bas

Bayou Penchant

Bayou Perot

Bayou Petit Liard

Bayou Petite Anse

Bayou Petite Calliou

Bayou Petite Passe

Bayou Petite Prairie

Bayou Pierre

Bayou Pigeon

Bayou Plaquemine

Bayou Plaquemine Brule

Bayou Point au Chien

Bayou Pointe aux Loups

Bayou Portage

Bayou Prairie

Bayou Queue de Tortue
Bayou Raphael
Bayou Rapides
Bayou Rapids
Bayou Rigolette
Bayou Rigolettes
Bayou Rixner
Bayou Rond Pompon
Bayou Roseau
Bayou Rouge
Bayou Sale
Bayou Salle
Bayou San Miguel
Bayou San Patricio
Bayou Santabarb
Bayou Sara
Bayou Sauvage
Bayou Scie
Bayou Segnette
Bayou Serpent
Bayou Shaffer
Bayou Sorrel
Bayou St. Denis
Bayou Stiff
Bayou Tambour
Bayou Tawpaw
Bayou Teche
Bayou Tent

Bayou Terre Blanc
Bayou Terrebonne
Bayou Tigre
Bayou Tommy
Bayou Tony
Bayou Toro
Bayou Tortue
Bayou Toulon
Bayou Tupawek
Bayou Vacherie
Bayou Verret
Bayou Wauksha
Bayou Wikoff
Bayou Wilson
Bayou Zourie
Bear Bayou
Beaver Bayou
Beck Bayou
Bee Bayou
Big Alabama Bayou
Big Bayou
Big Bayou Chene
Big Bayou Friejon
Big Bayou Pigeon
Big Carenero Bayou
Big Choctaw Bayou
Big Fork Bayou
Big Hog Bayou

Big Roaring Bayou

Big Saline Bayou

Black Bayou

Black Lake Bayou

Black Water Bayou

Blue Hammock Bayou

Bobtail Bayou

Bodcau Bayou

Boggy Bayou

Bossier Bayou

Bottle Bayou

Bourbeaux Bayou

Boutte Bayou

Bristow Bayou

Brushey Bayou

Brushy Bayou

Buffalo Bayou

Bull Bayou

Buras Bayou

Bushneck Bayou

Cache-Cache Bayou

Camper's Bayou

Campti Bayou

Canal Bayou

Caney Bayou

Carencro Bayou

Chatman Bayou

Chauvin Bayou

Chico Bayou

Choctaw Bayou

Clark Bayou

Clarke Bayou

Clay Cut Bayou

Clear Bayou

Corney Bayou

Coswell Bayou

Cottonwood Bayou

Coushatta Bayou

Cow Bayou

Cowpen Bayou

Creole Bayou

Crew Bayou

Crooked Bayou

Cross Bayou

Cypress Bayou

Dalton Bayou

Dave's Bayou

Davis Bayou

Dead Man Bayou

Deep Bayou

Dixie Bayou

Dolet Bayou

Doyle Bayou

Duck Bayou

Dutch Bayou

Elbow Bayou

False Branch Bayou

Felice Bayou

Fifi Bayou

Filly Bayou

Flagon Bayou

Flat Boat Bayou

Fouse Bayou

Freshwater Bayou

Goat Bayou

Grand Bayou

Grand Bayou Blue

Grand Louis Bayou

Grassy Bayou

Grayson Bayou

Green Cane Bayou

Greens Bayou

Gum Bayou

Haha Bayou

Half Moon Bayou

Harpoon Bayou

Heath Bayou

Hibbs Bayou

Holmes Bayou

Honey Bayou

Hooppole Bayou

Indian Bayou

Joe Bayou

Joe's Bayou

John's Bayou

Johnson Bayou

Keatchie Bayou

Kelly Bayou

King George Bayou

Kisatchie Bayou

Lacassine Bayou

Lapine Bayou

Larto Bayou

Lee Bayou

Liberty Bayou

Lick Bayou

Little Alabama Bayou

Little Bayou

Little Bayou Boeuf

Little Bayou Bonne Idee

Little Bayou Chene

Little Bayou de L'Outre

Little Bayou Long

Little Bayou Pierre

Little Bayou Pigeon

Little Bayou San Miguel

Little Bayou Sara

Little Bayou Sorrel

Little Bear Creek

Little Choctaw Bayou

Little Colewa Bayou

Little Corney Bayou

Little Cypress Bayou

Little Hog Bayou

Little Kisatchie Bayou

Little Mosquito Bayou

Little Tensas Bayou

Little Wallace Bayou

Little Wax Bayou

Locust Bayou

Loggy Bayou

Lonewa Bayou

Long Bayou

Long Island Bayou

Lost Dog Bayou

Marsh Bayou

Mayous Bayou

Mays Bayou

Middle Bayou

Mile Point Bayou

Mink Bayou

Mississippi Bayou

Mitchell Bayou

Monahan Bayou

Mothiglam Bayou

Mound Bayou

Muddy Bayou

Mundy Bayou

Oil Mine Bayou

Old East Bayou

Old North Bayou

Owl Bayou

Palmetto Bayou

Panola Bayou

Patout Bayou

Patrick Bayou

Petticoat Bayou

Pierre Bayou

Piney Bayou

Piquant Bayou

Plumb Bayou

Prairie Bayou

Pump Bayou

Rambin Bayou

Red Chute Bayou

Republican Bayou

Ross Bayou

Roundaway Bayou

Saline Bayou

Sam's Bayou

Sand Beach Bayou

Sandy Bayou

Sawdust Bend Bayou

Sawyer Pond Bayou

Shell Bank Bayou

Shoe Bayou

Sims Bayou

Siphorien Bayou